CHINESE RED
中国红

传统家具
Traditional Chinese Furniture

顾杨 ◎ 编著

全国百佳图书出版单位
时代出版传媒股份有限公司
黄山书社

图书在版编目(CIP)数据

传统家具：汉英对照 /顾杨编著. ——合肥：黄山书社，2012.8
（中国红）
ISBN 978-7-5461-3083-5

Ⅰ.①传… Ⅱ.①顾… Ⅲ.①家具—介绍—中国—古代 Ⅳ.①TS666.202

中国版本图书馆CIP数据核字(2012)第198739号

传统家具
CHUAN TONG JIA JU

顾 杨 编著

出 版 人：任耕耘
责任编辑：侯 雷　　　　　　　　　　　特约编辑：朱昌爱
责任印制：戚 帅 李 磊　　　　　　　　装帧设计：商子庄

出版发行：时代出版传媒股份有限公司（http://www.press-mart.com）
　　　　　黄山书社（http://www.hsbook.cn）
　　　　　（合肥市蜀山区翡翠路1118号出版传媒广场7层　邮编：230071）
经　销：新华书店　　　　　　　　　　营销电话：0551-3533762　3533768
印　刷：合肥精艺印刷有限公司　　　　电　话：0551-4859368

开　本：710×875　1/16　　　　　　　印张：10.75　　字数：137千字
版　次：2012年8月第1版　　2012年8月第1次印刷
书　号：ISBN 978-7-5461-3083-5　　　　定价：59.00元

版权所有　侵权必究
（本版图书凡印刷、装订错误可及时向承印厂调换）

前言 Preface

中国传统家具具有极高的历史价值和文化价值。数千年来，传统家具见证了中国古人的生活取向和审美情趣的变化，将雅与俗汇于一炉，将多种文化高度融合，具有强烈的民族风格和时代特征，与中国古典建筑及室内装饰风格一脉相承，在世界家具史上占有独一无二的地位。

Traditional Chinese furniture boasts a high historical and cultural value. For the past thousands of years, traditional furniture has been a witness to the changes in lifestyle and aesthetic taste of past generations. Suitable for both refined and popular appreciation, traditional furniture incorporates various cultures and powerfully features the national style and characteristics of the times. It shares the same origins of classical Chinese architecture and interior decoration and plays a unique role in the world history of furniture.

随着古人居坐方式从席地坐到垂足坐的演进，中国传统家具式样由低矮型发展到高型，种类从单一发展到多样，制作工艺愈加精湛高超……那深厚的文化底蕴、天然古朴的制作材料、合理的结构部件、简练典雅的造型、严密的榫卯结构、精雕细琢的装饰工艺等，无不令人惊叹。

本书介绍了中国传统家具的历史、制作工艺、装饰工艺、经典式样等，让读者在欣赏中国传统家具的艺术魅力的同时，进一步感悟中国传统文化的精深、优美。

As people's ways of life changed from cross-legged seating to chair-level seating (seating on chairs with feet suspended vertically above the ground), low traditional Chinese furniture was replaced with high furniture, with enriched varieties and greatly improved techniques. Its rich cultural meaning, natural and unsophisticated materials, logically arranged structural parts, simple and elegant design, seamless mortise-and-tenon joints and exquisite decorating techniques are all reasons to celebrate the genre.

Through an introduction of its history, making and decorating techniques and classical examples, this book presents the artistic glamour of traditional Chinese furniture to readers while broadening their knowledge about traditional Chinese culture.

目 录 Contents

传统家具概说
Overview 001

家具的由来和发展
Origin and Development 002

传统家具的辉煌
Glory of Traditional Chinese Furniture 020

中国传统文化与家具
Traditional Chinese Culture and Furniture 036

巧妙的制作技艺
Ingenious Manufacturing Techniques 043

传统家具的制作过程
Manufacturing Process of Traditional

Chinese Furniture 044

独特的榫卯结构
Unique Mortise-and-tenon Frame

Connection 056

传统家具的主要结构部件
Major Structural Members 065

精美的装饰工艺
Exquisite Decorative Crafts 089

雕刻装饰
Carving Decorations 090

镶嵌装饰
Inlay Decorations 103

漆饰
Painting Decorations 109

经典家具式样鉴赏
Classic Furniture Appreciation 117

床榻类
Beds and Couches 118

椅凳类
Chairs and Stools 124

桌案类
Tables and Desks 133

箱柜类
Cabinets and Chests 143

屏风类
Screens 151

支架类
Stands 157

传统家具概说
Overview

中国传统文化博大精深，如璀璨繁星，传统家具便是其中的一颗明星。中国传统家具历史悠久，自成体系，具有强烈的民族风格和浓厚的东方韵味，以其富有美感的永恒魅力引来众多中外人士的钟爱和追求。

Rich and sophisticated traditional Chinese culture is like a conglomeration of shining stars, one among which is traditional furniture. In the course of its time-honored history, traditional Chinese furniture has developed a system of its own featuring ,a strong national style and oriental appeal, attracting hundreds and thousands of people from home and abroad to pursue its everlasting glamour and beauty.

> 家具的由来和发展

中国传统家具历史悠久，源远流长。早在原始社会，先民们就用劳动的双手创造了家具。发展到明清时代，传统家具的发展完全成熟。

家具缘起

早在旧石器时代晚期，生活在华夏大地上的人类就掌握了结草成

● 余姚河姆渡遗址出土的漆木碗（新石器时代）
Lacquer Wood Bowl Excavated at Hemudu Relics (Neolithic Age)

> Origin and Development

Traditional Chinese furniture has a time-honored history. In the early years of society, Chinese progenitors made furniture using their bare hands. Furniture making was brought to maturity in the Ming Dynasty and Qing Dynasty.

Origin

In the Paleolithic Age, people living on the Chinese subcontinent already knew how to weave grass into mats for sitting or lying, or as bedding, marking the earliest known Chinese furniture. Among the Hemudu relics in Yuyao, Zhejiang province, during the Neolithic Age (about 7,000 years ago), grass mat objects were excavated, testifying to the fact that grass mats back then used herringbone and interlacing weaving techniques. Large bamboo mats and mats made of thin bamboo strips unearthed

● 仰韶文化遗址的房屋建筑示意图（新石器时代）
Housing Construction Sketch Map of the Yangshao Culture Relics (Neolithic Age)

席的技术。编织而成的草席可供人们坐卧铺垫，这是中国最古老的家具。从距今约7千年前的浙江余姚河姆渡新石器时代的遗址中发现的一些编织席的实物，可知当时的席就已采用二经二纬的"人"字形和交互编织工艺；在距今约五千年的浙江吴兴县钱山漾遗址中出土了大幅竹席和篾席，编织方法多种多样，其中还有较为复杂的竹编工艺。

among Qianshanyang relics in Wuxing, Zhejiang province, (about 5,000 years ago) were created using various weaving techniques, including some complex bamboo weaving methods.

Large numbers of stilt-style architectural relics were also discovered at Hemudu relics. Stilt-style architecture was normally located near a river with hills to the rear. They were built from frames made with wooden uprights or bamboo stakes on which there were

除此之外，河姆渡遗址中还发现了大量的干栏式建筑（由桩木、板桩、圆木组成的长排式建筑，上架设大、小梁（龙骨）承托地板，构成架空的建筑基座，建筑构件用榫卯连接。干栏式建筑一般依山而建，背山面水布置）。许多建筑木构件上已采用带有销钉孔的榫等结构形式，为制造家具提供了技术基础。在后来的家具结构和装饰中通常会借鉴当时流行的建筑木作技术。

large or small beams (also known as dragon bones) supporting a suspended floor. All the parts were connected by mortise and tenon joints. Of the unearthed relics, many timber pieces had structural forms employing pin-hole tenons, laying a technical foundation for furniture making. Subsequent furniture structures and decorations often drew on the popular timber making techniques of that time.

• 余姚河姆渡遗址出土的木构件
（新石器时代）
Timber Pieces Excavated at Hemudu Relics (Neolithic Age)

商周家具

商代（前1600—前1046）、西周（前1046—前771）和春秋战国时期诸国（前770—前221）都是在黄河流域建立的奴隶制王朝。这一时期，冶炼青铜的技术发达，奴隶主用青铜制作兵器、车辆、家具等，

Furniture in the Shang Dynasty and Zhou Dynasty

The Shang Dynasty (1600B.C.-1046 B.C.) and Western Zhou Dynasty (1046 B.C.-771 B.C.) and the Spring and Autumn and the Warring States Period (770 B.C.-221 B.C.) are all slavery dynasties built along the Yellow River. In this era, advanced

- **铜案（战国）**
此案造型简洁，案面中部镂空的勾云纹饰玲珑精巧、简练。
Bronze Table with Recessed Legs (Warring States Period, 475 B.C.-221 B.C.)
This recessed-leg table adopts a simple mould. The geometric motifs of openwork carving are exquisite.

- **云纹漆凭几（战国）**
此几每边各有三根足，呈并列状，均衡而对称。几面用一块整木雕成，浅刻云纹，两端雕刻兽面纹，精美生动。
Lacquer Side Table with Cloud Design (Warring States Period, 475 B.C.-221 B.C.)
This table has three legs arranged in parallel on either side, balanced and symmetrical. The surface plate is made out of a single piece of wood with a lightly carved cloud design and beautiful and lively animal face motifs on either side.

- **透雕漆木禁（战国）**

此木禁由整块厚木雕成，禁面阴刻云纹加朱绘，面板当中有一个"十"字隔梁。腿部是四只形象生动的野兽，兽的前腿向上弯曲，连接禁面与禁座，后腿环抱方柱。通身黑漆为底，朱绘花纹，有草叶纹、陶纹、鳞纹和涡纹等。

Lacquer Wood Jin of Openwork Carving (Warring States Period, 475 B.C.-221 B.C.)

This table (Jin) is carved out of a complete chunk of thick wood. Its face is carved in intaglio with cloud design and red paint. In the middle of the face is a criss-cross diaphragm. Its legs are in the shape of four vivid beasts, whose front legs curve upward, connecting the board and pedestal. The hind legs encircle the square column. The whole body is painted black and red-lacquered motifs, including grass leaves, earthenware, scales and swirls.

形成了独特的青铜文化。青铜家具主要是有足有座的青铜礼器，以方体四足和方形高座最为典型。这一时期家具的用途主要是展示器物拥有者的财富和地位，仅在祭祀、礼仪和大型宴饮场合才使用，大小有序、成套成列地摆放在一起。青铜礼器上的兽足、高座、加盖、附耳是当时的人追求美和高雅华贵的体现，这是当时家具的一大特点。

　　西周时，出现了髹漆技术和在漆木家具上镶嵌蚌壳的技术。但此时的家具制作总体水平不高，且加工木材的工具有限，所以漆木家具较少。

bronze smelting techniques enabled slave owners to make bronze objects such as weapons, chariots and furniture, thus forming the unique bronze culture. Bronze furniture was mostly bronze ritual vessels with legs and pedestals typified by square bodies with four legs or with a high pedestal. Furniture in this era was mainly used to showcase the fortune and status echniques enabled slave owners to make bronze objects such as weapons, chariots and furniture, thus forming the unique bronze culture. Bronze furniture was mostly bronze ritual vessels with legs and pedestals typified by square bodies with four legs or with a high pedestal. Furniture in this era was mainly used to showcase

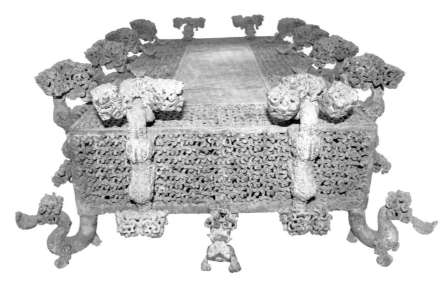

- **多层云纹铜禁（春秋）**

  禁是中国古代用来承放酒杯的案。此禁通高28.8厘米，通长131厘米，通宽67.6厘米，重90多公斤，呈长方形，类似于后代的几案，四周以透雕的多层云纹做装饰，禁身的上部攀附着12条龙形怪兽，探首吐舌，面向禁中心，形成群龙拱卫的场面，十分壮观。

  Multi-layer Bronze Jin with Cloud Design (Spring and Autumn Period, 770 B.C.-476 B.C.)

  Jin refers to tables on which ancient Chinese would put their wine glasses. This Jin measures 28.8 cm in height, 131 cm in length and 67.6 cm in width, and weights more than 90 kg. It is rectangular and looks like tables in later times. Around it is decoration of multi-layer openwork cloud design. Its upper part has 12 dragon-like beasts attached with their tongues sticking out and heads facing the center, forming a gorgeous spectacle of dragon guards.

- **凹形石俎（商）**

  俎是中国古代祭祀时放祭品的器物，多以青铜或石制成。

  Concave Stone Altar Stand (Shang Dynasty, 1600B.C.-1046 B.C)

  Altar stands are stands made of stone or bronze on which sacrificial food is put.

- 漆木座屏（战国）

此座屏线条曲美，雕工细致，描画生动。在这样小小的长条木框中，透雕和浮雕着数十只动物。

Lacquer Wood Screen Set in a Stand (Warring States Period, 475 B.C.-221 B.C.)

This screen features lovely curves, exquisite carving and vivid drawing and painting. Within this small wooden frame there are dozens of animals carved either in openwork or in relief.

- 朱绘漆几（战国）

此几由三块木板榫接而成，竖立的两块木板上端向内侧圆卷，下端平齐；面板两端有榫头，插入立板。

Red-lacquered Side Table (Warring States Period, 475 B.C.-221 B.C.)

This table is tenoned with three wood boards. The upper ends of the two vertical boards terminate in inwardly curly circles and the lower ends are parallel with the ground. The horizontal plate has tenon joints that fit in the vertical plates at both ends.

- 彩绘带立板漆俎（战国）

此俎面呈长条形，两端俎面之上有两块立板，立板及足板外侧均雕有纹饰。

Lacquer Altar Stand With Vertical Plates (Warring States Period, 475 B.C.-221 B.C.)

This elongated altar stand has two plates erected on its surface. Decoration motifs are carved both in the vertical plates and the exterior of the pedal plate.

春秋战国时期，木工作为一个行业出现了，由于席地而坐的居坐方式仍然盛行，因此这一时期主要是各式低矮的家具，包括席、床、榻、俎、禁、几、案、箱、笥等。

the fortune and status of its owners and only used in rituals, ceremonies and large banquets where sets of large and small furniture were arranged together in order. Animal legs, a high pedestal, lid and ears on bronze ritual vessels manifest the pursuit of beauty and elegance of the time and are typical of furniture in this era.

In the Western Zhou Dynasty, lacquering and clamshell-inlaying came into existence. Despite that, lacquered wood furniture was rare due to the backwardness of furniture making and lack of timber processing tools.

In the Spring and Autumn and the Warring States Period, carpenters became available on an industrial scale. Since cross-legged seating was still fashionable in this era, most of the furniture was made to be low, including mats, beds, couch beds, altar stands, tables, side tables, desks, boxes and bamboo-plaited baskets or suitcases.

 **十五连盏铜灯架（战国）**

此铜灯架为树枝形，高达84.5厘米，枝上饰有龙、鸟、猴、人物等形象，灯座由三虎承托，上有三条翼龙，十五盏灯可聚可散。

Bronze 15-candles Stand (Warring States Period, 475 B.C.-221 B.C.)

Shaped like twigs, this stand measures 84.5 cm in height. Its twigs are decorated with carved dragon, bird, monkey and human being images. The pedestal is propped up by three tigers, on which there were three winged dragons. Each of the candle stands can be dismantled.

秦汉家具

秦（前221—前206）、汉（前206—公元220）是大一统的封建王朝，幅员辽阔，人口众多，手工业蓬勃发展，漆器制作工艺成为最重要的家具制作工艺。高档的家具和器物，莫不优先使用漆器制作。汉

Furniture in the Qin Dynasty and Han Dynasty

The Qin Dynasty (221 B.C.-206 B.C.) and Han Dynasty (206 B.C.-220 A.D.) were unified feudal dynasties within a vast area and containing a huge population. It was a period when handicraft flourished

- 《高逸图》中席地而坐的文人（唐）
Writers Sitting on the Ground in the *Gaoyi Paining* (Tang Dynasty, 618-907)

包缘：为防止席边缘散落，用丝麻、绢、锦等织物制作的包边。
Edge covering: fabrics such as silk, helm and brocade are used to cover the edge to prevent it from fraying.

席面：多用草、竹等材料编织而成。
Mat: mostly woven of materials such as grass or bamboo.

- 锦缘莞席（西汉）

莞席制作较为粗糙，一般铺在地上使用。此席通长220厘米，宽82厘米。

Guan Grass (schoenoplectus validus) Mat of Brocade Covered Edge (Western Han Dynasty, 206 B.C.-25 A.D.)

Guan grass mat is usually spread on the floor for use. This mat measures 220 cm in length and 82 cm in width.

代漆器作坊林立，分工细致，有素工（内胎）、髹工、上工（漆工）、黄涂工（在铜制饰品上鎏金）、画工（描绘）等众多工种，且每道工序都由负责的工匠来署名。

这一时期，席地而坐仍是人们居坐的主要方式，供席地居坐的低矮型漆木家具进入全盛时期，并形

and lacquering techniques were of overriding importance in furniture making. High-end furniture and vessels were primarily made of lacquer. Lacquer workshops stood in great numbers in the Han Dynasty, in which there were detailed divisions of work, such as *Su Gong* (lacquer body), *Xiu Gong* (lacquer), *Shang Gong* (painting), *Huang Tu Gong*

成了较完整的系列家具。低矮型漆木家具的制作使用榫卯构造，装饰以彩绘为主，颜色由黑红彩绘发展到多彩，并出现了堆漆、镶嵌等新的装饰手法。这一时期，人们崇尚神仙，因此流云纹成为家具装饰最为显著的特征。

东汉时（25—220），西域文化借佛教东行而传入，人们的居坐习惯开始发生了变化，席地而坐开始向居坐床榻转变。家具的品种和样式也有了较大的发展，出现了由矮型家具向高型家具演变的端倪。

(gilding decorative bronze objects) and *Hua Gong*(drawing). Each procedure was signed by the craftsmen in charge of it.

In this period, cross-legged seating remained common. Low lacquered furniture for this purpose reached its peak with relatively complete sets of furniture series being made. Low lacquered furniture used mortise-and-tenon joints. The widely employed decoration developed from black and red to more colorful decoration and new techniques such as embossed lacquering and inlaying appeared. As people yearned for the celestial life, the flowing cloud pattern

- **云龙纹漆屏风（西汉）**

  此屏风是为随葬而制作的冥器，通高62厘米，长方形，屏板下有一对足座加以承托。屏板正面红漆地，绘有一条巨龙穿梭在云层里，边框饰朱色菱形图案。屏板背面用朱地彩绘几何方连纹，以浅绿色油彩绘，中心部分绘有谷纹璧。

  Lacquer Screen with Cloud and Dragon Design (Western Han Dynasty, 206 B.C.-25 A.D.)

  This screen is a rectangular burial object with a height of 62 cm. Under the panel there is a pair of foot pedestals to prop the screen up. On the front, there is red lacquer with a painting of a dragon flying through clouds. The frame is decorated with a vermilion argyle design. On the back, there is vermilion lacquer with geometric block design that is painted light green. The central part is decorated with valley line design.

- 云纹漆案（西汉）

先秦两汉时期流行小型食案，低矮、轻巧，案面四周设有拦水线，防止食物汤水溢出，适合人们席地而坐时持案进食。此漆案是专用于放置食器的食案，长方形，四周起沿，平底，底部四角有矮足，案内饰以由髹红、黑漆组成的精美云纹图案，色彩明晰，线条流畅。

Lacquer Table with Cloud Design (Western Han Dynasty, 206 B.C.-25 A.D.)

In the pre-Qin dynasties and the Han Dynasty, small dining tables were popular. They were low and light tables with water-barring skintle around the borders to prevent food and water from spilling. Such tables enabled people to eat their meals while sitting on the ground. This lacquer table was exclusively used for a dinner set. It is rectangular with high molding around the edge, a flat bottom, four short legs in four corners and an elaborate cloud motif of red and black lacquer characterized by clear color and fluent lines.

became the most distinctive feature of the furniture decoration at that time.

In the Eastern Han Dynasty (25-220), the culture of the Western Regions spread into the territory by way of Buddhism. People's ways of living started to change from cross-legged seating to life on beds and couch beds. Along with this came the changes in the types and patterns of furniture, indicating the evolution from low to high.

中国古人的坐姿

中国古代，人们的正式坐姿是"跪坐"，即两膝着地，两脚的脚背朝下，臀部落在脚踵上。在"跪坐"时，如将臀部抬起，上身挺直，就称为"跽"，是将要站起身的准备姿势，也是对别人尊敬的表示。

How did the Chinese Sit in Ancient Times

In ancient China, the correct way to sit was on one's knees, i.e. two knees down on the ground, the dorsum of the feet facing down and buttocks resting on ankles. This kneeling position was also regarded as correct etiquette by the Chinese of the time because the position of the buttocks and straightening of the upper body are preparations for standing up and therefore it shows respect to others.

- 商代晚期妇好墓中出土的跪坐玉人

Jade Woman Kneeling Down, Unearthed From the Tomb of Lady Fuhao, Late Shang Dynasty (around 1200 B.C.).

魏晋南北朝家具

魏晋南北朝时期（220—589），西北少数民族来到中原，带来了战争，也带来了胡椅、胡床、矮椅子、矮方凳等家具。汉族当作坐具的床榻也逐渐增高、增大，有的床还加上了床顶和床帐。同时，古印度佛教在中国迅速传播，带来了高型家具，如椅、凳、墩等。人们席地而坐的居坐方式开始改变，高型

Furniture in the Wei, Jin and the Northern and Southern Dynasties

In the Wei, Jin and the Northern and Southern dynasties (220-589), northern and western ethnic minorities entered the Central Plain and brought with them wars as well as the furniture of northern barbarian tribes, such as chairs, beds, low chairs and low stools. The Han people's couch beds for sitting were also elongated and enlarged, some even with canopies

家具开始萌芽。不过，这一时期的高型家具和垂足而坐的生活习俗只在贵族和僧侣中流行。

魏晋南北朝时期的家具用材除漆木家具外，竹制家具和藤编家具等也给人们带来了新的审美情趣。家具装饰受到异域文化的影响，不局限于秦汉时期的神兽云纹，还出现了莲花纹、火焰纹、飞天纹等具有清秀风格的装饰纹样。

and curtains. Meanwhile, ancient Indian Buddhism spread rapidly, bringing high furniture such as chairs, benches and stools. Cross-legged seating seemed to be losing its dominance and high furniture started to gain influence, but only among aristocrats and in monasteries.

In addition to lacquer furniture, bamboo furniture and rattan furniture also appealed to people aesthetically in this period. Influenced by non-China culture, furniture decoration was not limited to mythical icon and cloud motifs descended from the Qin Dynasty (221B.C.-206B.C.) and Han Dynasty (206B.C.-220A.D.) but also included lotus, flame and flying Apsaras motifs.

- **《列女古贤图》漆画屏风（南北朝）**
  漆画屏风是封建上层社会享用的奢侈品，共五块，内容主要是表现帝王、将相、列女、孝子以及高人逸士的故事。

  Women of Virtue and Scholars of Talent Lacquer Painting Screen (Northern and Southern Dynasties, 386-589)

  Lacquer painting screens were luxuries for the feudal upper class. This screen has five panels, with each telling a story of emperors, generals and ministers of state, women of virtue, filial sons or writers and scholars.

• **银香案（唐）**

香案是参佛所用。此香案为纯银制成，造型简洁、大气，曲线丰满、动感明快，没有过多的繁杂装饰，表达出敬佛者的虔诚。

Silver Incense Table (Tang Dynasty, 618-907)

The incense table is for the worship of Buddha. This incense table is made of pure silver with simple and elaborate molding, dynamic curves and unsophisticated decoration, showing the piety of Buddha worshippers.

隋唐五代家具

隋唐五代时期（581—960），人们的起居方式由席地而坐向垂足而坐转化，此时矮型家具和高型家具并存。

唐朝（618—907）的开放政策使得许多外国商人和佛教僧侣在各大城市长期定居，他们的生活习俗和审美取向对汉文化产生了重大影响。唐代家具出现了许多新的品种和装饰纹样，体现了盛唐时代气势宏伟、富丽堂皇的风格特征。

此外，由于唐代大量兴建佛座

Furniture in the Sui Dynasty, Tang Dynasty and the Five Dynasties

The Sui Dynasty, Tang Dynasty and the Five Dynasties (581-960) featured the gradual replacement of cross-legged seating with chair-level seating and the coexistence of low and high furniture.

In the Tang Dynasty (618-907), the opening-up policy attracted many foreign merchants and Buddhist monks to settle permanently in big cities. Their lifestyles and aesthetic tastes had a profound impact on the Han culture. New types and decoration patterns appeared among Tang furniture,

和佛塔，在须弥座丰富多彩式样的启迪下，中原工匠，创造出有束腰结构的新家具。此后，束腰家具层出不穷，影响到床、榻、桌、椅、凳、几等各个家具门类。

五代时期（907—960），社会的审美风气发生了改变。朴素、简明、风骨替代了大唐的华贵、厚重与圆润。高型家具成为主流家具，越来越多的家具是为垂足坐而设计的。家具高足化，装饰减少，应用框架式结构，显得挺拔、简明，为以后宋代家具简练、质朴的风格做好了铺垫。

manifesting the magnificence and grandeur of the flourishing period.

Moreover, inspired by the colorful and rich patterns of booming Buddhist pedestals and pagodas built in the Tang Dynasty, Chinese craftsmen created new furniture with waisted structures. Later on, waisted furniture was made in large numbers, influencing beds, couch beds, tables, chairs, stools, side tables, etc.

During the Five Dynasties (907-960), the aesthetic tastes of the society changed. Instead of emphasizing sumptuousness and plumpness as they did in the Tang Dynasty, people admired simplicity and vigor of style. High furniture gained dominance and more and more furniture was designed for chair-level seating. Furniture with high legs, simplified decoration and frame structures looked tall, straight and unsophisticated, paving the way for simple and unadorned furniture in the Song Dynasty (960-1279).

• 敦煌壁画中的架子床（唐）
Canopy Bed in a Dunhuang Grotto Mural (Tang Dynasty, 618-907)

- **《韩熙载夜宴图》中的家具（五代）**

从图中可看出，五代时人们已完全摆脱了席地居坐的习俗，进入垂足居坐时期。

Furniture in the Picture of *Night Banquet Hosted by Han Xizai* (Five Dynasties, 907-960)

From this picture, we can see that people in the Five Dynasties had already abandoned cross-legged sitting and entered the era of chair-level seating.

宋元家具

宋代（960—1279），人们的居坐方式为垂足而坐，与此相应的高型家具得到迅速发展，步入成熟时期。宋代家具的种类、形制及室内陈设都有了新面貌，多采用梁柱式框架结构，结构轻巧。宋代人以朴素、简洁、实用、工整规范为美，宋代家具也因此形成了简洁实用的结构和清秀素雅的装饰风格。

元代是蒙古族建立的大一统封建王朝，存在不足百年，故元代家具没有形成新的特色。但元代家具仍有许多创新：如抽屉桌、罗锅

Furniture in the Song Dynasty and Yuan Dynasty

In the Song Dynasty (960-1279), people sat on chairs with their feet suspended above the ground. This period was accompanied by the rapid development and maturity of high furniture. Types and designs of furniture and interior arrangement in this period took on a new look. The light frame structure of beams and columns was in wide use. Song people valued austerity, simplicity, practicability and standardization and, correspondingly, Song furniture had simple and pragmatic structures in an elegant and pretty decorative style.

椅、展腿式桌的出现，霸王枨这一新结构的运用，多用云头、转珠、委角等线型作装饰等。尤其是髹漆、雕花、填嵌工艺的发展，显示了元代家具制作技术的成就，并为明代家具的辉煌奠定了基础。

卧女长方足镜架（宋）
Rectangular Mirror Stand Shaped Like a Lying Woman (Song Dynasty, 960-1279)

• **铜镜和银镜架（元）**
Bronze Mirror and Silver Mirror Stand (Yuan Dynasty, 1279-1368)

The Yuan Dynasty (1279-1368) was a unified feudal dynasty built by the Mongols. It lasted less than a century, a period not long enough to lead to new characteristics in Yuan furniture. Nevertheless, Yuan furniture was innovative in its own way. For example, there was the emergence of tables with drawers, humpbacked chairs and tables with extended legs, the utilization of the giant's arm brace and linear decoration such as cloud heads and rotary beads. Octangular design also came into fashion. The technical development of lacquering, carving and inlaying were significant achievements in furniture making in the Yuan Dynasty and laid a solid foundation for the glorious Ming furniture.

> 传统家具的辉煌

中式家具内容博大精深，历史源远流长，在世界家具史上独占鳌头。尤其是古朴典雅的明代家具和繁缛华丽的清代家具，创造了传统家具的辉煌。

古朴典雅的明代家具

明代隆庆初年（1567），海禁解除，允许私人进行海外贸易，中国商品得以大量出口，海外货物也源源不断地输入中国。这其中包括东南亚地区产出的紫檀木、花梨木等，这为明代家具制作提供了充足的物质条件。

同时，从明代中期开始，社会经济繁荣，社会上追求享乐之风盛行，营建新民居、私家园林和装修房屋也时尚。而这些建筑中都设有

> Glory of Traditional Chinese Furniture

With its profoundness, extensiveness and long history, Chinese furniture is unique in world furniture history. In particular, the simple and elegant furniture of the Ming Dynasty (1368-1644) and sophisticated and graceful furniture of the Qing Dynasty (1644-1911) represented the glory of traditional Chinese furniture.

Simple and Elegant Furniture in the Ming Dynasty

In 1567, the maritime embargo was lifted and individual merchants were allowed to conduct trade overseas, making it possible for large exports of Chinese commodities and imports of foreign goods, including imports from Southeast Asia such as rosewood (also known as red sandal wood) and ormosia henryi (*Huali wood*), providing sufficient material resources for Ming furniture-making.

● 黄花梨木小座屏（明）

此座屏的屏心由大理石制成，做工精妙，既实用又美观，极具艺术性。

Yellow Rosewood Screen Set in a Stand (Ming Dynasty, 1368-1644)

The central panel of this small screen is made of marble and is exceptionally well carved, practical, beautiful and artistic.

厅堂、卧室与书斋，也需要高档次的家具，因此出现了各种与建筑空间相适应的桌案类、凳椅类、床榻类、柜架类、小器件与陈设用品等家具。

明代家具用材讲究，造型大方、优美，比例合度，结构科学，重在表现木材本身的色泽和自然纹理，制作工艺精湛，装饰风格简洁，被誉为中式家具史上的高峰时期，是中华民族的精粹文化之一。

At the same time, from the mid-Ming period onward, social prosperity gave rise to a pursuit of enjoyment and the fashion of building new residential houses and private gardens as well as fitting them up. Those houses were often furnished with dining rooms, bedrooms and study rooms, which required top-grade furniture and thus led to the emergence of various corresponding furniture such as tables and desks, stools and chairs, cabinets and bookshelves, as well as household accessories.

The furniture features of the Ming Dynasty are carefully selected materials, inclusive and beautiful design, appropriate ratios and science-based structures. It highlights the natural luster and grain of timber, exquisite manufacturing techniques and simple decoration. Those qualities have rendered the Ming Dynasty the peak period in the history of Chinese furniture and made Ming furniture the quintessence of the Chinese people.

In choosing materials, Ming furniture worships simplicity and taps fully into the natural grain and luster of timber, mostly using unadorned fine wood such as rosewood, yellow rosewood, blackwood, wenge (*Jichimu*), mesua ferrea, phoebe -zhennan (*Nanmu*), and beech to depict the aesthetic taste of the educated Ming for simplicity

在用材上，明代家具崇尚质朴之风，充分运用木材的天然纹理和色泽，多选用紫檀、黄花梨、红木、花梨木、鸡翅木、铁力木、楠木、榉木等优质木材，不加修饰，体现出明代文人追求古朴雅致的审美趣味。

在造型上，明代家具各部件比例合度、协调，基本与人体各部的结构特征相适应，使用起来十分舒适。各部件线脚造型挺拔秀丽，挺而不僵，柔而不弱，刚柔相济，并注重线形的变化、直线和曲线对比、方和圆对比、横和直对比，形成简练、质朴、典雅、大方的装饰风格。

在做工上，明代家具结构严

and internal harmony.

In molding, Ming furniture arranges different parts in a proportionate and coordinated way basically in line with the structural characteristics of the different parts of a human body, making it comfortable to use. Architrave molding of all parts features a combination of hardness and softness with an emphasis on changes in lines and comparison of the straight and curved, the square and round and the horizontal and vertical, fashioning a simple, elegant and respectable decorative style.

In manufacture, Ming furniture is well-structured and mainly uses mortise and

- **黄花梨木折叠式镜台（明）**
  此镜台设计谨严，雕工精美，精选黄花梨制成，每块装板上均有深色花纹，是明代小件家具中的精品。支架铜镜的背板位于镜台上层边框内，能放平，也能支成约为60°的斜面。台座两开门，中设三具抽屉。

  Yellow Rosewood Folding Mirror Stand (Ming Dynasty, 1368-1644)
  This yellow rosewood mirror stand is very exquisitely carved, with each panel decorated with dark-colored motifs. It is an elaborate work of small articles of furniture from the Ming Dynasty. Located within the upper frame of the stand, the easel can be put down or be unfolded to form an oblique plane at an angle of 60 degrees. The pedestal has two doors with three drawers.

抽屉：两个抽屉以短柱相隔，浮雕螭纹，醒目饱满，正中的浮雕下垂云头，安装吊牌。抽屉下面装板和下层的闷仓隔开。

Drawers: two drawers are divided by a short post and decorated with the conspicuous motif of *Chi* (a legendary dragon) in relief carving. In the middle of each drawer are carved with cloud heads in relief and a metal hang tag. The lower panel of drawers is separated from the lower pantry.

橱面：平整光滑，无翘头。

Surface: level and smooth, with no everted flanges.

闷仓：设于抽屉下，仓内可以存放物品，只有拉出抽屉才能取出仓内物品。上面雕刻螭纹，螭纹尾部恣意卷转。

Pantry: located under the drawer, for storage of goods. To take things out, drawers must be pulled out. *Chi* decoration motif is carved in relief. Its tail is rolled randomly.

牙板：牙板底部和吊头牙子的底部平齐，整齐美观，上浮雕缠莲纹，延绵不断，寓意生生不息。

Spandrel: its bottom, in parallel with the bottom of the protruding end, is neat and pretty; scrolling lotus motif is carved in relief, meaning the continuous circle of life.

角牙：设在吊头下面，浮雕缠莲纹。

Conner spandrel: located under the protruding end and carved in relief with a scrolling lotus design.

吊牌：活动拉手，可以旋转，使用方便。

Metal pull: movable pull that can be rotated, very convenient.

- **雕花闷户橱**

  此闷户橱造型大方华美，雕工细致精良，是明代家具向清代家具转型的作品。

  Pantry Cabinet with Carved Patterns

  This cabinet features elaborate molding and fine carving and was made during the transition from Ming furniture to Qing furniture.

● 黄花梨木五足内卷香几（明）

在明代，富贵之家有在书房、卧室内焚香熏屋子的习惯，香几常摆在室内宽敞处。圆形香几体圆，一般不带有方向性，从各个角度看都能形成完美的整体，面面宜人观赏。

Yellow Rosewood Incense Stand with Five Cabriole Legs (Ming Dynasty, 1368-1644)

In the Ming Dynasty, rich families had the habit of burning incense in their study rooms and bedrooms. Incense stands were usually placed in the middle of the room. Round incense stands are characterized by a round body, cabriole legs, and a three-dimensional style so they can be appreciated from every angle.

谨，主要运用榫卯结构进行连接，极少用钉和胶，牢固美观，没有丝毫的造作，给人朴素、文雅之感。

在装饰上，为体现名贵木料优雅的色泽和纹理，不破坏家具的完整性和自然美，工匠们仅在椅子的靠背板、桌案的牙子等显眼的部位，雕刻一些简单的纹饰或镶嵌小块玉石等，使小面积的装饰在大面积的素面中起到画龙点睛的作用。此外还以简单实用的金属饰件进行装饰，色泽柔和，使家具熠熠生辉。

17—18 世纪时，明代家具风格还对西方的家具设计产生过较大的影响。西方家具史上两个高峰时期的家具风格——法国巴洛克式家具

tenon joints rather than nails and glue to connect the pieces. It is solid, beautiful and unpretentious, giving viewers a sense of simplicity and gracefulness.

In decoration, craftsmen tried to show the elegant luster and grain of the fine wood and not to destroy the intactness and natural beauty of furniture. They only cared simple decorative motifs or inlay small jade stones in the backboard of the chairs, spandrel of desks and tables or other conspicuous positions to achieve the effect of simple decorating lighting up an expanse of unadorned area. Sometimes, simple but pragmatic metal decorative objects featuring gentle colors and luster are used to make a piece of furniture stand out.

From the 17th century to the 18th century, Ming furniture's influence was also keenly felt in the design of Western furniture. Both representative styles in the

搭脑：两头高翘，圆雕卷云头式花纹。

Top rail: ends terminate in well carved upward turning with cloud head motifs.

挂牙：设于横梁与立柱垂直相交处，透雕花云纹。

Hanging spandrel: located at the joints of the horizontal rail and vertical post and carved in openwork with a cloud motif.

"米"字形架：有两层，每层架用三根横枨榫接而成，将六条腿连接起来。

"米" shaped stand: two layers, each tenoned with three horizontal posts that connect all six legs.

霸王枨：设在上层"米"字形架和腿柱的交接处，起加固架子和承重的作用。

Giant's arm brace: located at the joints of the upper "米" shaped stand and leg posts, giving extra stability to the stand.

斜枨：卷云纹饰斜枨设于横梁与立柱间，承接横梁的重量，使其不易脱落。

Spandrel: carved with curved cloud motifs and located between the horizontal rail and vertical post to prop up the rail and prevent it from breaking off.

巾架：立柱与盆架两腿足连做，用于悬挂手巾。

Towel rack: vertical posts connected with two legs of basin stand, used to hang towels.

面盆架：由六根圆材制成，后面两腿和巾架的立柱相连。

Basin stand: six round legs, with two at the back connected to vertical posts of the towel rack.

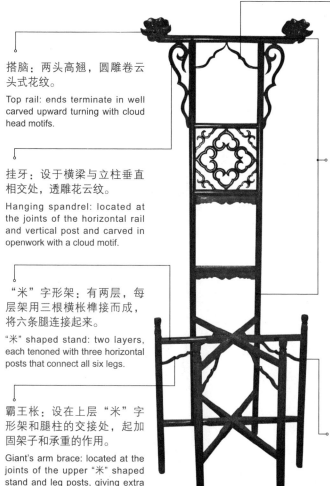

- **黄花梨木高面盆架（明）**

  此盆架制作十分精良，功用与装饰性能兼备，是一件传统家具精品。

  Yellow Rosewood Washstand (Ming Dynasty, 1368-1644)

  This washstand is exquisitely made and is a fine example of practical and beautifully-decorated traditional Chinese furniture.

和洛可可式家具，都从明代家具中受到过启发。明代家具弯曲优美的造型和线脚、装饰纹样、漆饰工艺已广泛用于西方新式家具中。

history of Western furniture— French Baroque and Rococo furniture— were partly inspired by Ming furniture to the extent that Ming curved and beautiful moldings, architraves, decorative motifs and lacquering techniques were widely employed in newly manufactured Western furniture.

- 固定式灯架（明）

固定式灯架是明清时期室内的主要照明用具，式样繁多，造型优美，亭亭玉立。

Fixed Lamp Stand (Ming Dynasty, 1368-1644)

Fixed lamp stands were widely used for interior illumination in the Ming Dynasty and Qing Dynasty (1644-1911). They have a rich variety and beautifully carved molding that looks like a tall dancing girl.

- 黄花梨木玫瑰椅（明）

明代黄花梨家具是具有强烈的文人风格的家具，不加雕饰，注重内涵的自然表露。此椅在椅背和扶手处营造出大片空白，表现出空灵之美和平淡的韵味，彰显了中国传统家具的美学特质。

Yellow Rosewood Rose Chair (Ming Dynasty, 1368-1644)

Rosewood furniture in the Ming Dynasty is of a strong scholarly style, unadorned yet naturally revealing its connotations. The blank space at the back and side rails of the chair expresses the beauty of emptiness and the appeal of moderation, showing the artistic beauty of traditional Chinese furniture.

苏州拙政园中卅六鸳鸯馆内的家具陈设
Furnishing in the Hall of the 36 Mandarin Ducks in Suzhou Humble Administrator's Garden

繁缛华丽的清代家具

清代康熙年间（1661—1722），传统家具进入了繁缛华丽的时代，造型厚重，形体庞大，雕饰繁缛，风格雍容华贵。清朝历代皇帝都对皇室家具的形制、用料、尺寸、雕饰、摆放等方面非常注重，并对此一一过问，使家具在造型和雕饰上尽显皇家的正统和威严。此外，当时贵族大量兴建私家园林，贵族之间斗奇夸富，侈靡之风盛行，反映在家具制作上是追求奢侈华贵的审美倾向。

清代家具大多用紫檀、花梨木、红木、楠木等名贵木料制成，质地坚硬，木纹美观。除了木质之外，清代家具还用象牙、大理石、宝石、珐琅、螺钿、竹藤、丝绳等多种材料，结合雕刻、镶嵌、漆饰

Luxuriant Furniture in the Qing Dynasty

During the reign of Emperor Kangxi in the Qing Dynasty (1661-1722), traditional furniture entered a phase of sophistication and gracefulness, featuring large and heavy designs, sophisticated decoration and elegant and poised style. Qing emperors attached great importance to the mould, material, size, decoration and arrangement of imperial furniture and

- **太师椅（清）**
  清代太师椅后背通常做成屏风式，扶手两侧有站牙（下部嵌入家具而垂直状的牙板），刻雕或镶嵌大理石、玉片、瓷片等作为装饰。
  Taishi Chair (Qing Dynasty, 1644-1911)
  Taishi (senior grand tutor) chairs in the Qing Dynasty feature a screen-type back, arms with standing spandrels on both sides and decoration of carved or inlaid marble, jade or pieces of porcelain.

嵌大理石：围屏中镶嵌具有成型花纹的大理石，形成一幅天然的水墨山水画。

Marble inlay: marble with decorative motifs inlaid in the three panels, forming an ink landscape painting.

暗八仙纹：由八仙纹派生而来流行于整个清代的宗教纹样，具有祝颂长寿、驱邪保平安的寓意。暗八仙纹中不出现人物，而以八仙所持之物代表各位神仙。

Invisible eight-immortal design: a religious design popular throughout the Qing Dynasty that is derived from the eight-mortal design. As a symbol of longevity and safety, invisible eight-immortal design features the objects held by the eight immortals rather than the immortals themselves.

- **酸枝木藤面罗汉床（清）**

清代罗汉床体积大，一般陈设在厅堂上。此罗汉床的床围为三屏式（由三块组成），正中稍高，两侧依次递减，镶嵌有大理石。

Blackwood *Luohan* Bed with Cane-woven Surface (Qing Dynasty, 1644-1911)

Luohan beds in the Qing Dynasty were large and usually placed in the hall. This bed features a railing with a three-panel screen: the panel in the middle is higher than panels on both sides and all have marble inlay.

"龙凤呈祥"纹:"龙凤呈祥"纹是一种有喜庆色彩的吉兆纹。多用于宫廷家具上,象征皇帝和皇后的和谐。民间家具上的龙凤纹,龙为草龙,凤为草凤,与宫廷所用龙凤纹不同,多用来贺新婚或祝颂夫妇生活和美。

Dragon-and-phoenix motif: a festive auspicious motif, mainly used on furniture belonging to a royal family, symbolizing the harmony between the emperor and the empress. By contrast, dragon-and-phoenix motifs on furniture belonging to ordinary families featured a straw dragon and a straw phoenix and were mainly used by the newly wed or to wish a couple a happy and harmonious life.

- **紫檀木五屏式镜台(清)**

  此镜台制作工艺精湛,外形秀美,每屏上横梁两端均挑出,雕饰花纹。镜台平面光洁,四周设有围栏。下部有对门开矮柜,可存放物品,柜门上饰有精致的铜饰件。

  Rosewood Five-panel Screen Mirror Stand (Qing Dynasty, 1644-1911)

  This mirror stand manifests exquisite workmanship and an elegant outer appearance. The side railings of each panel of the screen have protruding ends carved with decorative motifs. The surface of the mirror stand is bright and clean, surrounded by railings. On the lower part there is a low cabinet of double doors decorated with delicate bronze objects, where goods can be stored.

- 雕云龙纹书桌（清）

清代书桌的桌面多为长方形，尺寸一般不大，在结构上采用桌式结构，有几个抽屉的书桌已经十分普遍。特别是清代中期以后，受西洋家具的影响，出现了写字台式的书桌。其中尺寸特大者采用了类似架几案式的结构，做成可装配式的三件套，其下还备有踏板。

Writing Table Carved with Cloud-and-Dragon Design (Qing Dynasty, 1644-1911)
Most Qing writing tables are characterized by a small-sized rectangular surface and a table structure. Tables with drawers were very common. In particular, after the mid-Qing Dynasty, tables of a writing-desk structure came into existence under the influence of Western furniture, among which large ones adopted a structure similar to that of trestle tables, with 3 pieces in a set which can be assembled and a pedal in the lower part.

等多种工艺手法，精雕细作，并吸收外来文化的长处，仿制西欧家具样式。

清代家具制作技术远远超过前代，达到了炉火纯青的程度。木雕运用透雕和半透雕的手法，表现出空灵逼真的艺术效果。雕刻的纹饰图案丰富多彩，动物纹、人物纹、植物纹、器物纹、景物纹、几何纹等无所不有。宫廷贵族的家具多用"双龙戏珠"、"五福捧寿"、

would make their best efforts to make sure that furniture showed imperial orthodoxy and dignity in molding and decorating. In addition, noblemen in this period built numerous private gardens and vied with one another in wealth. This and the pursuit of extravagance are seen as an inclination toward deluxe design in furniture making.

Qing furniture is mainly made of hard fine wood with beautiful grains such as rosewood, blackwood and Phoebe zhennan (*Nanmu*). Besides wood, Qing furniture also uses other materials including ivory,

"祥云捧日"、"洪福齐天"等，表达出宫廷贵族炫耀权利和财富的功利性需求；老百姓的家具则多用"年年有余"、"百果丰硕"、"连生贵子"、"花开富贵"、"鹿鹤同春"、"凤穿牡丹"等，传达出民间老百姓朴素的生活意愿。镶嵌工艺，如嵌珐琅、嵌骨、嵌瓷、嵌螺钿、嵌玉石、百宝嵌等，在家具装饰上大放异彩。清代的漆饰达到了顶峰，家具中的大漆家具异彩纷呈，有大漆嵌百宝、戗

marble, gemstone, enamel, mother-of-pearl inlay, bamboo strip and silk cord. It integrates various techniques including carving, inlaying and lacquering. It draws upon the merits of foreign culture, imitating the furniture styles of Western Europe.

Furniture making techniques in the Qing Dynasty were far more developed than previous dynasties. Wood carving employed openwork carving and semi-openwork carving to achieve the artistic effect of vividness and lifelikeness. Decorative motifs were further diversified

- **雕狮纹嵌大理石面月牙桌（清）**
  月牙桌是桌面为半圆形的桌子，像一弯新月，委婉怡人。两张半圆桌可拼合成一圆桌，分开时可以单独靠墙放置，扩大了室内空间的利用率。清代月牙桌多见四足，富于雕饰，雍容华贵。

  Half-moon Table with Lion Design and Marble Surface (Qing Dynasty, 1644-1911)
  Half-moon tables have semicircular surfaces that resemble a half moon. Two semicircular tables can be put together to form a round table or separated and placed against a wall, thus extending the utility of interior space. A majority of Qing half-moon tables feature an elegant and luxurious style with four legs and sophisticated decoration.

- **榆木红漆描金柜（清）**

  此柜髹红漆，上有仕女图案描金装饰。仕女图案是以中国古代女性为表现内容的一种传统装饰图案。明清时期的家具上广泛运用仕女图案，婀娜多姿的仕女形象为家具增添了不少风采。

  Elm Red Lacquer Cabinet Painted with Gold (Qing Dynasty, 1644-1911)

  This cabinet is characterized by red lacquer and decorated with a picture of ladies painted in gold. The lady motif is a traditional decoration featuring traditional Chinese women. It was widely employed on Ming and Qing furniture. The pretty and charming image of ladies greatly enhances the elegance of the furniture.

金银、蜡饰、漆绘、黑漆描金、红漆描金等，富丽堂皇。

此外，清代家具在风格上形成了不同的地域性特色，以苏州、广

into various kinds such as animal, human being, plant, vessel, landscape and geometric motifs. The furniture of aristocrats largely adopted images such as "two dragons playing with a pearl", "five blessings and longevity", "propitious clouds holding the sun" and "limitless blessing", showing aristocrats' need to show off their power and fortune. By contrast, the furniture of ordinary families mainly used animals and plants images to convey their desire to lead a simple and happy life. Such as "having extra fish", "harvest of rich fruit", "bearing a succession of sons", "flowers blossom for riches", "deer and crane celebrating spring together" and "phoenix and peony". Inlaying techniques such as inlaying of enamel, bone, porcelain, mother-of-pearl, jade and multiple precious objects emerged widely in furniture decoration. Lacquer decoration reached its peak in the Qing Dynasty, and lacquer furniture stood out among all others, including lacquer inlaid with multiple precious objects, lacquer with gold and silver relief, wax on lacquer, lacquer painting, black lacquer with gold design and red lacquer with gold design, rendering Qing furniture magnificent and splendid.

州、北京为代表，形成"苏式"、"广式"和"京式"等不同风格的家具流派，标志着清代家具进入一个新的时期。

Moreover, Qing furniture featured different styles in different localities typified by Suzhou, Guangzhou and Beijing, leading to the formation of schools of different furniture styles such as the Suzhou style, Guangzhou style and Beijing style. This signified Qing furniture's entry into a new era.

- 鸡翅木座屏（清）

此屏风端庄典雅，呈现出鸡翅木栗褐色的色泽、绚丽的纹理。屏心镶嵌天然大理石，自然的石纹形成一幅极具水墨趣味的山水画；屏心和屏框之间透雕出空灵精美的几何图案；上下两块屏板分别浮雕博古纹和"福禄寿"纹；牙板透雕灵芝纹和团寿纹，两侧挂牙透雕灵芝纹；底座圆雕蝙蝠、寿字等福寿如意的纹饰。

Wenge Screen Set in a Stand (Qing Dynasty, 1644-1911)

This screen is elegant and graceful, revealing wenge's chestnut color and gorgeous grain. The central panel is inlaid with natural marble. Its grain forms a landscape painting with the use of ink. Between the central panel and the frame an openwork exquisite geometric motif is carved. The three gods of happiness, luck and longevity are carved on the two screen panels in relief, employing the design of antique patterns. On the spandrel is openwork carving with ganoderma lucidum design and *Tuan Shou* design (in Chinese, *Tuan* means reunion and *Shou* means longevity). On the hanging spandrel is also the openwork carving with ganoderma. On the pedestal is an openwork carving with designs that symbolize happiness, longevity and good luck, such as a bat and the Chinese character "寿" (*Shou*).

● 北京北海公园内的家具陈设
Furnishing in Beijing Beihai Park

> 中国传统文化与家具

传统家具作为传承中国文化的重要载体，在几千年的发展历程中，深深地烙上了宗教哲学思想、建筑、书法、绘画等传统文化的印记。"天人合一"、"顺应自然"等哲学思想，以及建筑、书法、绘画潜移默化地影响着家具设计的理念，造就了博大精深、辉煌灿烂的传统家具文化。

"天人合一"的自然观

传统家具融汇了中国传统的哲学思想，即儒、释、道三种宗教文化思想的精髓。

儒家文化主张"天人合一"，讲究天、地、人是一个整体，和谐一致。传统家具色泽幽雅而不沉闷，肌理华美而不艳俗，正体现了

> Traditional Chinese Culture and Furniture

As an important carrier of traditional Chinese culture, traditional furniture has reflected such things as religious and philosophical thought, architecture, calligraphy and painting in the course of its several thousand years of development. Philosophical ideas of "harmony between man and nature" and "following the laws of nature" as well as architecture, calligraphy and painting have been influencing the design philosophy of furniture, thus shaping the profound and splendid traditional furniture culture.

"Harmony between Man and Nature"

Traditional Chinese furniture absorbs traditional Chinese philosophical thoughts, namely the quintessence of Confucianism, Zen Buddhism and Taoism.

Confucianists uphold the idea of "harmony between man and nature",

这种"天人合一"的和谐性，给人美的感受。比例尺度严密，圆中有方、方中见圆的家具设计理念，也是儒家天圆地方哲学思想的体现。

道家思想主张"道法自然"，崇尚自然、含蓄、冲淡、质朴的天然美，提倡人与自然之间的契合无间。制作家具的木材，生长于自然，人们将其应用于家具、建筑及日常生活中时，也力求顺应自然，与自然相和谐。传统家具中的佳作，既要考虑木的自然特性，又要体现木家具的独特风格，因此都做得自然，毫不做作；结构、线条顺势而下，展现出木材优美、自然的纹理；纹饰源于自然界的动物、植物、人物、器物、景物等图案，也

regarding heaven, earth and man as a harmonious whole. The traditional furniture that features elegance and bright colors and luster plus extravagance instead of pretentious textures echoes such a Confucianist harmony, giving people a experience of beauty. The perfect proportion and the design philosophy of square in circle and circle in square also reflect the Confucianist philosophical thought that heaven is round and the Earth is square.

Taoism advocates that "the Tao (Way) follows nature", emphasizing the natural beauty of simplicity, spontaneity and detachment as well as integration of the individual and the cosmos. Wood for furniture making grows in nature and, when it is used in furniture making, construction and daily life, people try to follow laws of and harmonize with nature. To produce fine pieces of traditional

- **黄花梨木小方凳（明）**
此凳用浅色黄花梨制成，造型简洁，不加雕饰，经过细致的打磨上蜡，焕发出清丽、圆润的光泽，流露出自然、华贵的美感。
Yellow Rosewood Square Stool (Ming Dynasty, 1368-1644)
This stool is made of light-colored yellow rosewood and its molding is simple and unadorned. It is exquisitely polished and waxed, displaying a fresh and mellow luster and a natural and elegant beauty.

体现出自然之美。

　　禅宗崇尚人与自然浑然如一，主张明心见性，清除各种杂念，追求心灵的平衡和宁静，回归生命本身。禅宗思想简淡、素雅、自然、单纯、俭朴，然而简素中展现着臻于极致的完美。传统家具中有很多家具，尤其是书房家具，式样简洁、素朴，却通过边角、线条等细节部位实现了富丽的效果，流露出浓浓的禅意。此外，传统家

Chinese furniture, craftsmen not only consider the natural characteristics of wood but also seek to show the unique style of wood furniture. As such, exquisite furniture is unaffected: its structure and lines make good use of the natural beauty and grain of wood and its decorative motifs that originate from natural icons such as animals, plants, human beings, vessels and landscapes also manifest the beauty of nature.

Zen Buddhism values the unity of man and nature and believes in finding one's true self, removing distracting thoughts, pursuing equilibrium and tranquility of the soul and returning to life itself. Zen thoughts are simple, natural and pure but amid simplicity and purity is near-perfection. Most traditional furniture, especially furniture in study rooms, features simple and unadorned

- **榉木官帽椅（明）**
  此椅造型柔婉，弯转弧度大，搭脑、靠背板、后腿、扶手等皆是弯曲的，线条柔和、舒展，富有弹性，符合人体工学设计原理，与人体各部分的比例相协调，使人体能保持自然放松的姿态。

  Official Hat-shaped Beech Armchair (Ming Dynasty, 1368-1644)
  This chair features a soft and pretty molding, a large curve radian, curved top rail, back, rear legs and arms, and soft, stretching and flexible lines. Its design is ergonomics and the proportion coordinates with that of the human body so that people sitting on it will find it relaxing.

- **天然木根边座百宝嵌座屏（清）**

  此座屏的边座用天然木根制作，借助木根天然的形态、纹理、节疤、色泽，做成屏风底座和边框，与屏心的百宝嵌《九老观画图》极为匹配，给人以古香古色之感。

  Screen with Hundreds of Precious-Material Inlay (Qing Dynasty, 1644-1911)

  This screen is set in a stand made of natural root. It takes advantage of the root's natural form, grain, knot and luster to make the screen pedestal and frame, thus matching them with the hundreds of precious-material inlay *Nine Elder Men Appreciating a Painting* in the central panel, hence the antique flavor.

具中那些富于哲理性的几何、文字纹饰，不但起到了画龙点睛的装饰效果，而且还体现了禅宗以少胜多的意趣。

传统家具与书画艺术

　　书法是中国汉字的书写艺术，散发着浓浓的艺术魅力。书法作为中国传统文化的一个符号，其独特的审美原则和笔墨意趣直接影响着传统家具的造型。

　　书法佳作，往往笔画交接处婉转圆润，刚中带柔，字体苍润丰满。而传统家具亦是做工考究，具有浓郁的书法韵味。比如马蹄足家

styles, while elegance is achieved through attention to details such as corners and lines, a natural flowing of Zen philosophy. Furthermore, philosophical decoration of geometrical motifs and motifs of characters on traditional furniture not only mark the finishing point but also convey a Zen Buddhist joy in defeating the many with the few.

Traditional Furniture and Painting and Calligraphy

Calligraphy in China is an artistic way of writing Chinese characters with its own charm. As a symbol of traditional Chinese culture, calligraphy directly bears on the molding of traditional furniture with its

具，造型柔婉流畅，极像书法中的顿笔提钩。又如许多圆腿造型的家具在很多部位均锉成小弯弧，模仿书法的笔墨效果，使家具表现出挺劲丰满的力度美。

传统家具造型受中国绘画艺术的影响也颇深，其简洁劲秀的线条即是受中国画严谨、写实画风的影响。清代家具受明末清初画家崇尚写意画风的影响，很多家具出现了程式化、符号化的倾向，装饰性强，实用性降低，完全演变成为一种写意符号。

- 马蹄足

家具的足是腿的着地之处，足端通常有兽爪、马蹄、如意头、卷叶等样式。马蹄足的外形很像马蹄，同时也很像书法中的顿笔提钩。

Horse-hoof Foot

The feet of the furniture are where the legs stand, the end of which is usually shaped into animal claws, horse hooves, *Ruyi* ends, curling leaves and so on. Horse-hoof feet resemble horses' hooves from the outside and the pausing and hook-shaped stroke in calligraphy.

unique aesthetic ideologies and ink and painting temperament.

Excellent calligraphy work has been described as the art of giving form to signs in an expressive, harmonious and skillful manner. It is hard yet soft and involves vigorous and expressive characters. Traditional furniture involves the lingering charm of calligraphy through molding. For example, fluent modeling of the taut foot furniture calls to mind the pausing and hook-shaped stroke in calligraphy. Another example is that most round-legged furniture imitates ink writing and files different parts into small vertical hook strokes to manifest the forceful and vigorous beauty of strength.

Traditional furniture molding is also shaped by the art of Chinese painting featuring precise and realistic drawing, as a result of which lines in furniture are often simple yet vigorous. Influenced by the impressionistic style of painters in the late Ming Dynasty and early Qing Dynasty, Qing furniture was inclined to be stylized and metaphorical, characterized by an enhanced emphasis on decoration rather than practicability, creating a purely impressionistic symbol.

苏州留园内的家具摆设
Furnishing in Lingering Garden (Liu Yuan), Suzhou

- **紫檀木嵌木画座屏（清）**

  传统家具常以山水乡居、田园风光、树石花卉、庭院小景、亭台楼阁等山水画面做为题材的装饰纹样。山水纹的图案多取自历代名人画稿，意境典雅清新，陈设性强，常被装饰在屏风、柜门、柜身两侧和箱面、桌案面等大面积的面板上。

  Rosewood Screen Set in a Stand With Wood Painting Inlay (Ming Dynasty, 1368-1644)

  Traditional furniture often features decorative motifs of landscapes such as mountains and rivers, rural scenes, trees, rocks and flowers, courtyards, as well as pavilions, terraces and towers. Mountain and river motif often imitates famous paintings, which is elegant, refreshing and decorative. It is commonly used as decoration for screens, cabinet doors and cabinet sides, as well as large panels such as box and table surfaces.

巧妙的制作技艺
Ingenious Manufacturing Techniques

中国传统的木工技艺一直是被世人所称道的。木工运用巧妙的构思，创造出传奇般的榫卯结构，并且设计出了各种精巧的结构部件，从而发展出千变万化的家具。

Traditional woodworking techniques in China have long been praised by people throughout the world. Using their ingenious designs, carpenters have created the legendary mortise-and-tenon frame and made numerous elaborately structured pieces, without which the development of a veritable kaleidoscope of furniture would have been impossible.

> 传统家具的制作过程

传统家具的制作可分为十几个工艺流程,即选料、配料、画线、开料、木部件细加工、开榫凿眼、雕花、磨活、攒活、净活、打蜡擦亮等。虽然在实际制作时,工序的划分可能比本文所说还分得细致些,但具体的操作都包括在其中了。

选料、配料

选料和配料是根据家具的形制选择合适的木材。一般要选择木材的纹理(便于刮刨,也兼顾器表的美观)、颜色(配料)、用料的尺度大小。

画线

在木材上画出加工的记号。

开料和部件细加工

开料就是通过锯割,将板材加工成枋形毛料;之后用刨子加工,

> Manufacturing Process of Traditional Chinese Furniture

Traditional furniture making involves such technical processes as choice of material and dosing, line drawing, saw cutting, detailed processing of timber pieces, mortising and tenoning, carving, polishing, assembling, dressing and waxing. Although in reality, there maybe a more detailed version as regards technical processes, the above list is already extensive.

Choice of material and dosing

This refers to the choosing of wood suitable to the furniture structure. Generally, elements to be considered are grains of wood (for easy scraping and the beauty of the furniture's surface), color (for dosing) and size of the wood materials.

Line drawing

Drawing marks where processing

将枋形毛料加工成符合标准形状、标准尺度的精枋枋形料；最后，便要画线，按家具各个部位连接的情况，画出所需的榫卯结构。

开榫凿眼

即按画线位置，精确做出榫、卯结构。做榫头是用锯割的方法，做卯眼是用凿子凿卯眼。虽然加工

• 古书中的木工工作场景
Carpenters at Work in Ancient Books

is needed.

Saw cutting and detailed processing of timber pieces: saw cutting is to use a saw to cut wood into shaped raw materials. A plane is then used to process shaped raw materials into refined, standardized shaped pieces. The final step is line drawing, namely drawing lines where mortising and tenoning are needed according to how different parts are connected.

Mortising and tenoning

Precise making of the mortise-and-tenon frame connection in accordance with the lines drawn. Tenons are cut with a saw and mortises are chiseled. Even when taking great care, precise mortising and tenoning is hard to achieve and dressing tools are necessary to make numerous corrections in order to ensure the wood materials are angled correctly and mortise-and-tenon joints are seamless.

Trial assembling

Attempt to assemble pieces of timber with ready-made mortises and tenons into independent structural units, mainly for checking whether the joints are of suitable sizes and seamless and if there is any tilting or imperfect angle. Adjustments should be promptly made to

过程十分小心，榫、卯总是会有不妥当之处，要用修整工具对榫头和卯眼进行多次细心修正，保证装配后的木料是相互垂直的，榫口连接处是严丝合缝的。

认榫

即将做好榫卯的木部件，试组装成一个相对独立的结构单元，主要是检查榫卯是否大小合适、是否严密，有无歪斜或翘角等情况。如发现不妥，要及时修整，确保每一个结构部件单

ensure that the surface of every structural unit fits perfectly.

Wood carving

Work on furniture pieces such as motif carving and applying architraves. This technical process is conducted after the trial assembling, when all parts are carefully disassembled and those that need carving and having architraves attached are processed. It is so arranged because only after jointed timber pieces are correctly shaped and planed can

• 古书中的木工工作场景
Carpenters at Work in Ancient Books

元的表面都符合严格的尺度规定。

雕花

即在家具部件上进行雕刻花纹和起线等工作。这道工序要安排在试组装之后，把各种部件轻轻拆开，在需要进行雕花起线的部件上进行加工。雕刻纹饰之所以要在认榫之后进行，是因为连接在一起的木配件已定形，确保在同一平面之上，这时进行雕花起线可确保所有的雕花浅深一致。如违反这个次序，则不能保证雕花部件都在同一平面上。当然，有些雕刻是采用整板嵌入结构。雕花是一个大工序，其中分为画活、雕刻、做细等工序步骤。

磨活

指在未组装之前将每个木部件进行打磨，使每一个部件都达到表面平整精细，无刀痕、擦痕等。传统工艺是用泡湿的锉草捆成草把，将各个部件的每个表面都仔细打磨几遍；再用泡湿的光叶（冬笋的外皮）顺着纹理仔细打磨，所以行业术语叫"磨活"。经过磨活的木材表面非常光滑，用手抚摸感觉不到任何的凹凸不平，也看不见刻痕和横向的擦痕。现代改用水砂纸和机

wood carving ensure a consistent depth. If the sequence is reversed, carved pieces may not be on a level plane. Of course, sometimes, wood carving is completed by inlaying a whole plate. Carving is a demanding process that can be further divided into technical steps known as drawing, carving and refining.

Polishing

In this process, timber pieces are polished to remove tool marks and scratches so as to appear level and refined. Traditional techniques use bundles of common scouring rush herbs to polish the surface of every piece and then further polished with wet leaves from the outer skin of winter bamboo shoots along the grain. The polished wood surface is very smooth without any tangible roughness or visible notches and transverse scrub marks. Today's polishing is accomplished with waterproof abrasive paper and machinery. Efficient and effective as it is, it still lags behind traditional polishing materials in smoothness and moistness.

Assembling

Assembling means to assemble all parts. It is also known as "fish gluing". Independent structural units, such as doors and windows, panels and flanks

鳔

虽然传统家具的榫卯做得确实很严实，但家具在组装时仍然要"用鳔"，即用动物胶把木材的榫卯粘固。

鱼鳔和猪皮鳔都是动物胶，优点是在常温下会冻结，有很强的黏性，受热时又会恢复为溶液而失去黏性。正因为这个优点，所以中国古代家具的榫卯都用动物胶粘合，也便于维修，但也因此有了怕水浸、怕受热的弱点。

Fish Glue

Although the mortise-and-tenon joint in traditional furniture is firm enough, furniture still needs "fish glue" during assembly to reinforce the connection.

Fish glue and pigskin glue are two types of animal glues. They freeze under normal temperature and have strong viscosity. They melt, becoming less viscous if heated. These qualities enable them to be widely used for gluing tenons and mortises in traditional Chinese furniture. This makes furniture easy to maintain but vulnerable to water and heat.

械打磨，虽然工效很高，打磨效果也很好，但终不如用传统磨光材料那样圆润。

攒活

攒活是把所有的部件正式组装起来，也叫"使鳔"。一般分立的结构单元，如门扇、面板、侧山要先行组装，经过测量没有尺寸上的误差，待干透后，方可进行全柜的组装。正式攒活时，把各种部件备齐，按次序摆放好，在鱼鳔热好后，分别在榫头和卯眼中涂上热鱼鳔；装好后用布擦

need to be assembled and measured first to ensure the preciseness of their sizes before they dry and are ready to be assembled into a whole. To complete the assembly, prepare all the parts and arrange them in order. Then coat tenons and mortises with hot fish glue, removing the extruded fish glue with a towel after assembling. At this moment while the fish glue is still hot, check the preciseness with a ruler so that minor errors such as unevenness can be promptly corrected by pressing and pulling. If no errors are found, leave the assembled furniture

去挤出来的鱼鳔。这时要趁鱼鳔未凉之前，迅速用尺子校验装配的精度，如有不方、不正的小误差，可用挤压推拉的方式及时调整。如装配无误，便可静置一两天，等鱼鳔胶自然干透。

净活

指对组装好的家具做最后的修整。一件家具组装好后，要静置一两天，等鳔干透后才能进行净活。净活的工作内容是对木材接口处微小不平之处，用耪刨进行刮平修整，之后还要对新加工处打磨干净，把胶迹刮擦干净等，以进行染色和烫蜡。

火燎

火燎指对组装好的家具做最后的修整。如果各方面检查无误，还要对白茬（因此时的家具呈灰白色，故名）家具进行火燎处理，即用酒精（古代用高度白酒）均匀涂在家具上，然后点燃。目的是利用酒精燃烧将家具表面翘起的细小木刺烧掉（现代工艺不用火燎，改用热水擦和水砂纸打磨）。这样做可保证在染色后，家具的表面依然平整细腻，保证后序烫蜡擦亮的质量。

behind for one or two days to let the fish glue dry naturally.

Dressing

The final conditioning for assembled furniture. Assembled furniture should be put aside for one or two days until the fish glue dries. After that, the furniture is dressed. Dressing mainly involves removing minor uneven areas around joints with a centipede plane, polishing the newly processed areas and scraping away glue marks before dyeing and waxing.

Singeing

The final dressing for assembled furniture. If all parts are checked and found to be precise, any white stubble needs singeing by daubing it with alcohol (strong Chinese white wine was used in ancient China) and then burning it. The purpose is to burn off wood splinters on the furniture surface (modern techniques do not involve singeing. Instead, hot water is used for scraping and waterproof abrasive paper for polishing). This is to ensure the smoothness and evenness of the furniture surface after dyeing to guarantee the quality of the follow-up waxing.

Dyeing

The preparatory work before waxing.

刷色

对家具进行烫蜡之前的准备工作。刷色是用棉丝或软布蘸上泡好的染色剂，顺木纹均匀擦拭，要避免染色剂流淌。干燥后视颜色深浅及木色的情况，再进行二三次染色。染色要施用一定的手法，才能保证家具上的颜色完全均匀一致，特别是成套的家具。硬木家具一般都要染成红褐色，所以俗称"红木家具"。传统染色剂是用热酒精浸泡紫檀木的粉末，浸泡数天之后，再经过滤，即成。

烫蜡擦亮

这是传统家具最后一道工序，即用加热法把石蜡熔融在家具表面上，待其部分渗入木表层后，及时用柔软的白布用力擦磨多次，使家具表面显示出美丽的木纹、颜色和光泽。不同木材的颜色和光泽是不相同的，像紫檀家具呈深黑色，红木家具呈红色，黄花梨家具呈琥珀色等。

Dyeing involves dipping cotton silk or mull in ready-made stains and then wiping them along the grain without allowing any of the liquid to drip. After the dye is dry, the process is repeated two or three more times, depending on the shade of the stain and the color of the wood. Dyeing requires skills to make sure that the color of the furniture, in particular if it is part of a set, matches perfectly. Hard wood furniture is usually dyed mahogany, hence the name "redwood furniture". Traditional stains are made by immersing rosewood powders into alcohol for several days. The mixture is then filtered.

Waxing

The final process in making traditional furniture. It is carried out by coating the surface of furniture with melted wax, waiting until some of the liquid has penetrated the surface and then using white mull to scour, ultimately revealing the beautiful wood grain, color and luster. Different woods boast different colors and luster. For example, rosewood furniture becomes black, redwood furniture red and yellow rosewood furniture amber.

现代家庭中的传统家具摆设

Modern Pieces of Traditional Furniture

传统家具的制作工具

在没有现代化机械设备的农耕社会，传统家具的制作仅凭简单的木工工具，征服了硬度很高、难以锯割的木材，制作成了千姿百态的家具。

度量画线工具包括直尺、弯尺、规、绳、折尺、活尺、勒子等；砍削工具包括斧、锛等；锯割工具包括框锯、刀锯、镂弓子等；刨平工具包括大刨子、二虎头刨、光刨、一字刨、凸刨、凹刨、平槽刨、槽刨、耪刨、铲刨和边刨等；雕刻工具包括扁铲、圆口、和尚头、齐口等；钻孔工具包括拉钻、舞钻等。

Tools for Traditional Furniture Making

In the agrarian society where no modernized machinery was available, traditional furniture making depended on simple woodworking tools to work the extremely hard wood that is difficult to cut and saw into furniture of various forms.

Traditional tools include measuring and marking tools such as a straight ruler, a curved ruler, compasses, rope, a folding ruler, a movable ruler and cutting and chopping tools such as an axe and adz; sawing tools including a frame saw, a saber saw and a fretsaw; A jack plane, a carpenter plane (used for removing raw materials), a smoothing plane, a horizontal plane, a convex plane, a fluting plane, a router plane, a rabbet plane, a loosening plane, a shoveling plane and a rebate plane are also needed. Carving tools such as a flat spade, round tongs and front cutting pliers also form part of the equipment. To complete the set, hole drilling tools such as *La Zuan* (literal meaning: pull drill) and *Wu Zuan* (literal meaning: dance drill) are required.

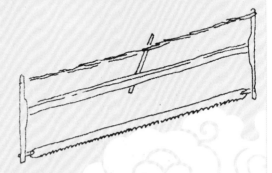

- **框锯示意图**

  框锯的种类较多，有横断锯和纵断锯之分（锯齿的角度不同），还有锯条很窄的"挖锯"。

  Frame Saw

  Frame saw boasts a large variety, such as crosscut saw and rip saw (depending on angles of saw teeth), and "dig saw" with narrow saw blade.

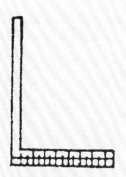

- 角尺示意图

  角尺是古代木匠画方形或直角线的工具，也用来校验刨平后的板、枋材以及结构之间是否垂直和边棱成直角。

  ### Set Square

  A set square is a tool used by carpenters of old to draw square or right angle lines. It is also used to check whether the planed plate, square log and other structural timbers are vertical and whether their uprights are at a right angle.

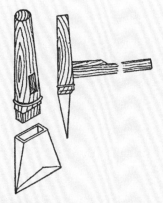

- 锛示意图

  锛的刀刃与木柄是垂直的，这与斧头不同，使用方法如锄头。锛用于大圆木料的削平，工效比斧头高。

  ### Adz

  Different from axe, blade of adz is perpendicular to its wood handle and to use it is like using a hoe. Axe is used to bevel round timber and is more efficient than axe.

- 墨斗示意图

  墨斗是弹画长直线时所用的工具。

  ### Carpenter's Ink Marker

  It is a tool used to draw long straight lines.

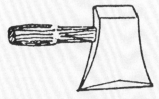

- 斧示意图

  常用的木工斧有双刃斧和单刃斧。双刃斧的刀刃在中间，宜做粗加工。单刃斧的刀刃在一侧，宜做细加工。

  ### Axe

  Types of axes commonly used by carpenters are double-bladed axes and single-bladed axes. Blade of double axe is in the middle and is suitable for rough processing whereas blade of single axe is on one side and is suitable for refined processing.

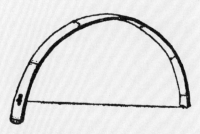

- 锼弓子示意图

锼弓子是用厚竹片弯成弓形，两端钻孔，绷装钢丝而成。钢丝上剁出刺状飞棱，利用飞棱利刃来进行锯削。钢丝极细，是用于镂空和曲线形加工的专用工具，凡有镂空花纹的牙子都要用锼弓子进行加工。另外，家具上常有曲线形的部件，为了节省材料，常采用套料排法，也要用到锼弓子。

Fretsaw

Fretsaw is made by bending thick pieces of bamboo into an arch, drilling a hole at each end and then furnishing tight steel wire. Cut spinous flying shuttles on the steel wire and those shuttles are used for sawing and cutting. Because of the thinness of steel wire, fretsaw is a special tool for openwork carving and curvilinear processing. For example, it is used to process spandrels with motifs carved in openwork. In addition, furniture usually has some curvilinear members. To save materials, trepanning drilling is a good and common choice, which also requires the employment of fretsaw.

- 拉钻示意图

拉钻由握把、钻杆、拉杆和牵绳等组成。钻杆长500毫米，直径35毫米，上端部制有圆榫，以便与握把配合；握把长80-100毫米，直径与钻杆相同，内有圆孔，用两瓦状竹片与钻杆顶部圆榫相接，可自由转动；拉杆长600-700毫米，直径约15毫米，两端有固定牵绳用的孔。使用时，左手拿住握把，右手拉动拉杆，事先缠绕在钻杆上的牵绳被拉动后，牵绳与钻杆之间产生摩擦力；随着拉杆的往复拉动，使钻杆作往复转动而起到钻孔的作用。

- *La Zuan*

La Zuan is kind of drill which composed of a grip, drill rod, pull rod and guy rope. The drill rod is 500mm long and 35mm in diameter and there is a dowel to coordinate it with the grip on its top. The grip is 80-100mm in length and also 35mm in diameter. Inside it there is a circular hole. The grip is connected with the dowel on the top of drill rod using two tile-shaped bamboo pieces and can turn freely. The pull rod is 600-700mm long and 15mm in diameter. On either side of it is a hole for fastening the guy rope. To use *La Zuan*, one needs to hold the grip with the left hand, and pull the pull rod with the right hand. The rationale behind it is that when the guy rope preliminarily wound ariller.

- 光刨示意图

光刨用于去堑碴或找平。

Smoothing Plane

Used for removal of ballast or leveling.

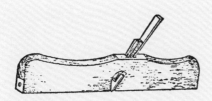

- 平槽刨示意图

平槽刨是用来刨削高低平槽及制作合角接合的工具，又称"杀肩刨"，刨底平直，刨刀比刨身略宽（不大于0.5mm）。

Router Plane

Router plane, also known as "shoulder-removing plane", is a tool used to plane and cut high and low flat-bottom slot and make convergence joints, whose bottom is flat and straight and cutter is a litter broader than adze block (no more than 0.5mm).

- 槽刨

槽刨是用来刨削板槽，以及装板时开小槽沟的专用工具。刨底有前后两条厚2mm、宽25mm铁片纵向嵌入刨身，铁片伸出8mm，构成直轨作为槽滑道。前后铁片中间为刨刀。刨身一边还需配上可调节之靠具。

Rabbet Plane

Rabbet plane is exclusively used to plane and cut board slot and dig small groove when inserting the plane. At the front and back of the plane bottom, there are two 2mm-thick and 25mm-wide iron sheets longitudinally inserted into adze block. The sheets stretch out by 8mm to form a straight rail that can be used as plane slide. In between the iron sheets is the plane cutter. On one side of adze block should also be equipped an adjustable leaning device.

- 铲刨和边刨

铲刨的刨形短小，边刨的刨形长大，构造相仿，都是用来刨削铲口、拼缝和制作合角接合的工具。

Shoveling Plane and Rebate Plane

Shoveling plane is short and small while rebate plane is long and big. They are similar in moulding and are both used for planning and cutting shovel mouth, piecing and making convergence joints.

- 一字刨示意图

刨底镶有铁板，这种刨子刨身短，接触面小，轻巧玲珑，使用灵活方便，适用于刨削不同形状的圆弧和弯曲工件。

Horizontal Plane

Inlaid with iron plate at the bottom, this kind of plane features short adze block and small contact area. It is light and flexible, and is used to plane and cut various forms of arc and curvilinear members.

＞ 独特的榫卯结构

中国传统家具的灵魂即榫卯结构。整套家具不使用一根铁钉，却能使用几百年甚至上千年，这在人类轻工制造史上堪称奇迹。

传统家具的榫卯结构即家具各个部件之间的连接方式。据统计，

＞ Unique Mortise-and-tenon Frame Connection

A mortise-and-tenon frame connection constitutes the soul of traditional Chinese furniture. It strikes a wonder in the light-manufacturing history of mankind by enabling a set of furniture with no iron nails to survive several hundred or even 1,000 years.

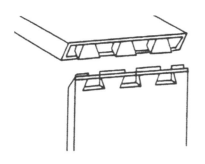

- 暗榫

暗榫即两块木板两端对接，接头形似燕尾，不外露，是制作几、案、箱子之类必用之榫。

Hidden Tenon

Hidden tenons refer to joints of two panels with swallow-tail ends. Hidden tenons are essential for making desks, tables and cases.

传统家具的榫卯结构近百种，常见的有栽榫、托角榫、粽角榫、长短榫、夹头榫、抱肩榫、楔钉榫、插肩榫、套榫、挂榫、勾挂榫、格肩、攒斗等。这些用榫卯结构于家具的不同部位，解决了传统家具的框架结构的美观性和牢固性问题。

传统家具的榫卯结构设计得非常科学，每一个榫头和卯眼都有明确的固定锁紧功能，能在整体装配时发挥作用。只要做工非常准确精细，榫卯之间是滑配合，略施一些鱼鳔，家具就非常结实牢固，而且

The mortise-and-tenon structure refers to the connection of all furniture parts. It is estimated that there are nearly 100 types of this structure, with the common ones known as planted tenon, prop corner tenon joint, triangular corner tenon joint, long-and-short tenon joint, elongated bridle joint, embracing-shoulder tenon, peg tenon joint, inserted shoulder joint, through tenon, hang tenon joint, hook-and-plug tenon joint, double-mitered tenon and latticework. These types of mortise-and-tenon structures are used in different parts of the furniture, forming beautiful and secure frame

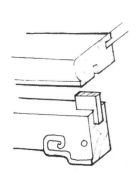

- **栽榫**

栽榫是一种用于可拆卸家具部件之间的榫卯结构。罗汉床围子与围子之间及侧面围子与床身之间，多用栽榫。

Planted Tenon

Planted tenon joints refer to a mortise and tenon connection applied between detachable furniture pieces, frequently found between the railings of a *Luohan* (a spiritual practitioner who has realized certain high stages of attainment in Buddhism) bed and between the side railings and the bed itself.

● 攒斗

攒斗指利用榫卯结构，将许多小木料拼成各种大面积的几何花纹，非常好看、非常结实，这种工艺叫"攒"。用镂镂而成的小木料，簇合成花纹叫"斗"。攒斗工艺原本是中国古建筑内檐装修制作门窗格子心、各种楠花罩的工艺技术，后来应用于家具制作中，一般制作架格的栏杆、床围子时用这种工艺。

Latticework

Latticework, also known as *Zan Dou* in Chinese, refers to large, pretty and sturdy pieces of geometric motifs formed by connecting small wood materials using the mortise and tenon connection. The technique used is called *"Zan"* (joining) and decoration motifs formed by small wood materials are called *"Dou"* (patches). Latticework was originally a technique in the interior furnishing work of traditional Chinese architecture for making door and window lattices and various carved covers. It was later applied to furniture making, usually used for making shelf and bed railings.

在家具的外表上根本看不见木材的横断面，只有凭借木材纹理的通断不同，方可看到榫卯的接缝。正是这些工艺精巧的榫卯结构，造就了传统家具的工艺特色。

connections in traditional pieces.

The structural design of a mortise-and-tenon is very scientific, with every joint having its own specific function. Precise woodworking, the coordination of mortises and tenons and the use of some fish glue are sufficient to make strong and firm furniture. Moreover, cross sections of wood materials are invisible from the outside and joints of mortises and tenons can only be seen through differences in wood grains. Exquisitely structured mortise and tenon joints create traditional furniture's characteristics of craftsmanship.

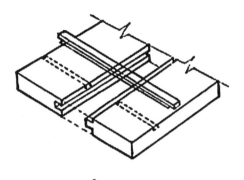

- 龙凤榫

龙凤榫是将数块薄木反拼成一块宽大的木板必用的工艺，一般要加穿带。其作法是将薄板分别"拼缝"，在第一块木板上的侧面做出一个银锭形（即断面为上窄下宽的梯形）的长榫，在第二块木板的侧面，凿出一个上窄下宽的梯形状的凹槽，再将第一块木板的长榫推入第二块木板的槽中，两块木板就严丝合缝地拼在一起了。

Dragon-and-phoenix Joint

This is a technique necessary for splicing several pieces of thin wood into a broad wood panel, usually with an additional penetrating transverse brace. The steps are: Using thin pieces of wood, make a long dovetailed tenon (trapezoid with narrow top and wide bottom) on the flank of the first panel, chisel a narrow-top wide-bottom trapezoid groove on the side of the second panel, push the long tenon into the groove so that the two panels are connected seamlessly.

- 燕尾榫

燕尾榫因榫头、卯眼的形状端部宽、根部窄，很像燕子的尾巴而得名，用于两块木板的连接，一般抽屉四周的木框、木箱、木提盒上用得最多。

Swallow-tail Tenon

It gets its name from its swallow-tail shape, featuring a tenon and mortise with a wide front end and a narrow root. It is mainly used to splice two panels, commonly found in the surrounding wood frame of drawers, wooden boxes and wooden hand-carried boxes.

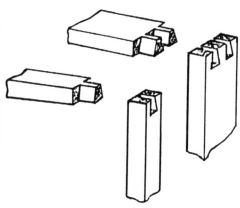

- 长短榫

一般腿部与面板的边抹接合时，腿料出榫做成一长一短互相垂直的两个榫头，分别与边抹的榫眼结合，故称"长短榫"。

Long-and-short Tenon Joint

For connecting legs with a panel Bian Mo, the leg material is crafted into two orthogonal tenons, one long and the other short, to match with the mortises of the Bian Mo, hence the "long-and-short tenon joint".

- 托角榫

托角榫即角牙与腿足和牙条相接合，一般在腿足上挖槽口，与角牙的榫舌相接合，当牙条或面子与腿足构成的同时，角牙与牙条或面子都打榫眼插入桩头，故"托角榫"是一组卯榫的组合，不是单一的构造形式。

Prop Corner Tenon Joint

Prop corner tenon joints refer to joints between corner spandrels and legs and aprons. Usually, a groove is dug in the legs to match the spandrels' tenon. In the connection of the apron or the outer side with a leg, both spandrel and apron or the outer side needs to be mortised and inserted with pile heads. So, "prop corner tenon joint" is a set of mortise and tenon joints rather than a single structural form.

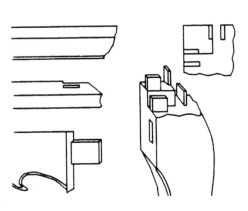

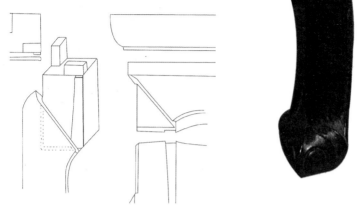

- 抱肩榫

抱肩榫是有束腰家具的腿足与束腰、牙条相结合时使用的榫卯结构。也可以说是家具水平部件和垂直部件相连接的榫卯结构。

Embracing-shoulder Tenon

It is a structural form used to join legs, the waist and aprons of waist furniture. It can also be described as a mortise and tenon structural form that connects vertical and horizontal parts of furniture.

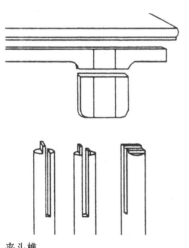

- 夹头榫

夹头榫是制作案类家具常用的榫卯结构。此种结构是利用四足把牙条夹住，连接成方框，上承案面，使案面和腿足的角度不易变动，并能很好地把案面板的重量分散，传递四条腿足上来。

Elongated Bridle Joint

It is a mortise and tenon structural form commonly used for tables. It utilizes the four legs to entrap aprons and forms a square frame. Its upper part supports the table surface, while evenly distributing the weight to four legs.

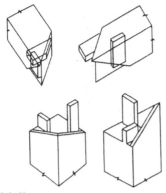

- 粽角榫

粽角榫因其外形像粽子角而得名，从三面看都集中到角线的是45度的斜线，多用于框形的连接。此外，明代家具中还有"四平式"桌，其腿足、牙条、面板的连接均要用粽角榫。

Triangular Corner Tenon Joint

Triangular corner tenon joints (*Zong Jiao* joint) derive their name from a resemblance to the corners of parcels of sticky rice wrapped in bamboo leaves which are eaten during the Dragon Boat Festival. Looking from three sides, one can only see oblique lines at an angle of 45 degrees. This joint is commonly used as a frame connection. It was also used to connect legs, aprons and panels in a table with straight flat sides during the Ming Dynasty (1368-1644).

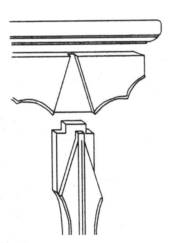

- 插肩榫

插肩榫是制作案类家具的榫卯结构。装配时，牙条与腿足之间是斜肩嵌入，形成平齐的表面；当面板承重时，牙板也受到压力，但可将压力通过腿足上斜肩传给四条腿足。

Inserted Shoulder Joint

The inserted shoulder joint is a mortise and tenon connection for table making. It is inserted during the assembly between the apron and legs, forming a level surface. When the surface panel bears a load, pressure is also exerted on the apron but the pressure is distributed among four legs through the sloping shoulder of legs.

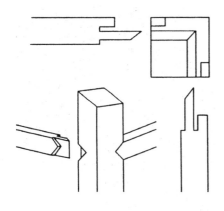 

- **格肩**

遇有横竖木料相交时,一般要将出榫料的外半部皮子截割成等腰三角形,另一根料上的半面皮子也制成同大的等腰三角形豁口,然后相接交合在一起,通称"格肩"。

Double-mitered Tenon

When horizontal and vertical wood materials meet, an isosceles triangle is usually cut in tenons and an isosceles triangular opening of the same size in mortises. Mortises and tenons are then connected together to form a mortise and tenon joint, generally known as a double-mitered tenon.

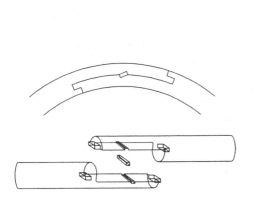 

- **楔钉榫**

楔钉榫是用来连接圆棍状又带弧形的家具部件,如圆形扶手的榫卯结构。

Peg Tenon Joint

The peg tenon joint is used to connect pole-shaped furniture parts with arcs, such as the round arms of a chair.

苏州退思园荫余堂的家具摆设
Furnishing in Yinyu Hall, Suzhou Retreat and Reflection (Tuisi) Garden

> 传统家具的主要结构部件

家具是由各种结构性部件，按照一定的工艺要求组合而成的。这些结构部件有各自的用途，制作时也有特定的要求。

面板

大多数家具都有面板，如桌面、案面、凳面、椅面等，一般用于承放物品。另外，柜子的柜门、侧山等结构性部件，虽不用于承重放物，但制作工艺也与面板相同。

中国传统家具一般不采用较厚的独木板来做面板，因为这样做有费材料、易变形、上下端有顶铆、木材难配等缺点。所以，大多数都采用"攒边框装板心"的方法来制作，即用四根规格相同的木条组成了一个方框。用于竖直方向的木条

> Major Structural Members

Furniture is made up of various structural pieces in accordance with certain technical requirements. Those pieces are used for different purposes and made under certain requirements.

Panel

Most furniture has a panel to put things on—for example, a table panel, a desk panel, a bench panel and a chair panel. In addition, structural parts of cabinets and chests such as doors and flanks also involve the same manufacturing techniques as panels, even though their uses differ.

Panels in traditional Chinese furniture rarely use single thick pieces of wood because it wastes material, is susceptible to distortion and it is hard to find wood that fits. As a result, most panels are made by joining the frame

叫"大边",用于水平方向的木条叫"抹头",大边与抹头的两端均采用45度角的对肩榫连接,便组成了一个方框。在"大边"和"抹头"的内侧,预先刨出一个小槽,将拼好的板心的"舌簧"部分插进槽内,便与木框构成了一个完整的平面。为了增加木框的强度和防止板心变形,在两条"大边"之间和面板心之下还加一二条"穿带"。

and inserting the panel core, i.e. forming a square frame by joining four battens of the same specifications. Vertical battens are called *"Da Bian"* (tenon-bearing frame member), and horizontal battens are called *"Mo Tou"* (window stool); ends of *"Da Bian"* and *"Mo Tou"* are connected by shoulder-to-shoulder tenon joints at an angle of 45 degrees, producing a square frame. Then a small groove is dug in the *"Da Bian"*

镶平面
Parallel inserting

大边
Da Bian

抹头
Mo Tou

- 方桌(清)
  Square Table (Qing Dynasty, 1644-1911)

嵌板心有三种不同的装法。一是镶平面，即板心与边抹攒合时处在同一水平面上，传统家具中的桌子、几案等多采用镶平面；二是"落膛踩鼓"，即将要装入的面板四周减薄，让中间高起约0.5厘米，然后嵌装在边抹围成的框架中；三

and "*Mo Tou*" and the reed part of the panel core is inserted into the groove to form a complete plane with the frame. To enhance the frame's firmness and prevent the panel core from deforming, one or two strips of "penetrating transverse brace" are added between two vertical battens and beneath the panel core.

There are three ways to insert the panel core: parallel inserting, i.e. keeping the panel and battens in parallel while joining them together, usually seen in traditional tables and desks; floating panel with raised center and recessed sides, i.e. thinning the periphery of the panel so that the middle part is about 0.5 cm thicker and then inserting the panel in the frame formed by battens; surface lowering inserting, i.e. intentionally

- 方脚柜（现代）
Cabinet with Square Feet (Modern Times)

落塘面
Surface lowering inserting

是落塘面，即将心板嵌入边抹围成的框架时，有意使板面低于框架，这种做法常用于柜子的门和侧山。

束腰

束腰家具的面板与支撑框架之间有缩进安装的牙条。束腰原本是佛教建筑中须弥座的上枭与下枭之间的部位，后来束腰作为一种结构

making the panel surface lower than the frame while inserting the panel core in the frame, usually seen in doors and flanks of cabinets and chests.

Waist

On waist furniture, an apron is installed between the panel and supporting frame. Waist originally referred to the part between the top and bottom caps in Buddhist buildings Sumeru Base and was later used as a structural element in furniture.

The earliest waist furniture appeared in the Wei, Jin, and Southern and Northern Dynasties (220-589) and was then passed down continuously. By the Ming Dynasty (1368-1644) and Qing Dynasty (1644-1911), waist furniture

- 榉木茶几（清）
Beech Tea Table (Qing Dynasty, 1644-1911)

束腰
Waist

- **榉木供桌（清）**

此桌为高束腰，桌面设栏杆，四周装透雕绦环板,牙子和三弯腿肩部均浮雕精美纹饰,足端向上翻卷,托泥攒接冰裂纹。

Beech Altar Table (Qing Dynasty, 1644-1911)

This table features a high waist, railings on the surface, surrounding ornamental panels in openwork carving, apron and squanders with elaborate decorative motifs in relief carving, upturning feet and continuous floor stretcher joining cracked ice design.

方式被移植到家具上。

有束腰的家具最早出现在魏晋南北朝时期，历代不断，至明清时更加流行，用于床、桌、凳、几、椅、柜等家具的面板与支撑框架之

was even more popular, commonly used between the panels and supporting frames of furniture such as beds, tables, stools, desks, chairs and cabinets. In terms of furniture structure, waist is noticeably decorative, makes the panel appear thick

须弥座

须弥座象征佛祖所居住的须弥山，有辟邪和吉祥如意的寓意。魏晋南北朝时期，佛教艺术大量传入中国，须弥座作为基座的一种式样被广泛用于神龛、坛、台、塔、幢及等级较高的建筑物之上。座身由许多条凹凸、宽窄不一的水平线脚组成，雕有各种纹样的图案如卷草、莲瓣、掐珠、壸门、动物、云水纹等，也具有很强的装饰性。

Sumeru Base

The Sumeru base mirrors Mount Meru where the Buddha lived. It is a symbol that is supposed to exorcise evil spirits and bring good luck and happiness. In the Wei, Jin and Southern and Northern dynasties (220-589), Buddhist art was largely spread to China and Sumeru base was widely used as a type of pedestal in shrines, altars, stands, towers, stone pillars engraved with Buddhist scriptures as well as some high-level buildings. The body of the base is made up of many concave and convex horizontal architraves of different width, plus carved decorative motifs such as curling tendrils, lotus petals, rotary beads, Buddhist doors, animals and cloud and water, which is very decorative.

- 北京故宫乾清宫前的须弥座
  A Sumeru Base in Front of the Palace of Heavenly Purity, Imperial Palace, Beijing

间。从家具的结构来看，束腰不仅使面板显得厚实又富有变化，有显著的装饰功能，还能增强面板和支撑框架的牢固度。

牙板

牙板又称"牙条"、"牙子"，一般用薄于边框的木板制成，安装在家具前面及两侧框架边沿，具有装饰和加固作用。牙板有素牙（无雕刻花纹）与花牙板（有雕刻花纹）之分，花牙板上雕有云纹、回

and solid and more diversified and can enhance the firmness between the panel and supporting frame.

Spandrel

Spandrel, also known as apron, is usually made of wood panels thinner than the frame panel and installed in the front of furniture and along the edges of the frames on either side. It serves as decoration and reinforcement. Spandrels are divided into uncarved spandrels and carved spandrels with carved motifs such as clouds, angular spirals and *Ruyi* (Chinese characters 如意 that bear the symbolic meaning of having all one's dreams come true). Some spandrels are named after where they are installed, such as the foot spandrel, corner spandrel, hanging spandrel, arch-shaped apron and gourd-shaped apron.

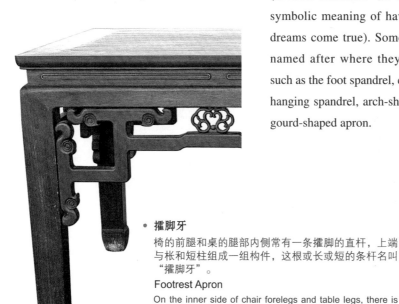

- **擢脚牙**

椅的前腿和桌的腿部内侧常有一条擢脚的直杆，上端与枨和短柱组成一组构件，这根或长或短的条杆名叫"擢脚牙"。

Footrest Apron

On the inner side of chair forelegs and table legs, there is usually a railing for resting feet. It forms a structural set with stretchers and short posts. This long or short railing is called a footrest apron.

- **券口牙**

  在大边与抹头组成的方框内侧，一般会安装四根牙条，主要起加固边框的作用，也有很好的装饰效果。

  Arch-shaped Apron

  Four aprons are usually installed inside the frame formed by *Da Bian* and *Mo Tou*, mainly serving to reinforce the frame, and are decorative.

- **角牙**

  角牙指安装在边框角部的短牙条。其中，有在家具下部垂直安装者，叫"站牙"；在家具靠上部位垂挂安装者，叫"挂牙"；呈倾斜状安装者，叫"披水牙"。

  Corner Spandrel

  Corner spandrels refer to short aprons installed in the frame corner, among which those vertically installed at the lower part of furniture are called standing spandrels. Those hung near the upper part are called hanging spandrels. Those obliquely installed are called slanted aprons.

- **挂牙**

  挂牙指上部嵌入家具而垂挂者，多见衣架、屏风类家具。

  Hanging Spandrel

  Hanging spandrels refer to spandrels whose upper part is inserted into furniture such as clothes racks and screens.

- 披水牙

披水牙是一种安装后呈斜坡状的牙条。最常见的披水牙是屏风两脚与屏座横档之间，两面均一条斜坡状安装的长条花牙。有些做工较好的条凳，其左右两条腿之间也有斜坡状安装的披水牙。

Slanted Apron

Slanted aprons are aprons that look like a slope after installation. The most common slanted apron is the long decorative apron located between the two legs of a screen and the horizontal panel of the screen stand, which is obliquely installed on both sides. It is also can be seen between the right and left legs of some well manufactured benches.

- 站牙

站牙是指衣架、灯架、屏风类家具下部的结构部件，一般是以垂直状态嵌入家具立柱的两侧，用来固定立柱。由于站牙常镂作葫芦之形，故又名"壶瓶牙子"。

Standing Spandrel

Standing spandrels are a structural item located at the lower part of such furniture as clothes racks, lamp stands and screens. They are usually inserted vertically into the two sides of the post for reinforcement. Due to the fact that standing spandrels are usually shaped like gourds in openwork carving, they are also called gourd-shaped apron.

纹、如意纹等。有些牙板因安装部位特殊而有专名，如脚牙、角牙、挂牙、券口牙、壶形站牙等。

托腮

托腮是位于束腰与牙子之间的一根木条，常做成挺括的线脚，具

Stepped Apron Molding

Stepped apron molding is a molding between the waist and the apron. It is usually a straight and smooth molding for decoration and reinforcement of the waist, either made from the same wood as the apron or a separate piece of wood.

托腮
Stepped apron molding

• 紫檀木雕云龙纹长方桌（清）
Rosewood Rectangular Table Carved with Cloud and Dragon Design (Qing Dynasty, 1644-1911)

有装饰和加固束腰的功能。有的与牙子一木连做，有的分做、另安。

枨

枨一般指除大边、抹头之外起加固作用的小木条，用料要比大边和抹头小一些。枨有多种形式，如

Stretcher

Stretchers often refer to small battens for reinforcement, excluding *Da Bian* and *Mo Tou* that use relatively larger battens. Stretchers have a large variety, such as the side stretcher, humpbacked stretcher, giant's arm brace, base stretcher and crossed stretcher.

横枨：横枨是一根用料较小的直木条，水平安装在桌、案、凳、椅的腿足之间；有的用一根，叫单枨，也有用两根，叫双枨。有的横枨有专名，如"管脚枨"、"裹腿做"。

Side stretcher: straight battens made from relatively small pieces of wood that are horizontally installed between the legs of tables, desks, stools and chairs. Side stretchers can use one batten, called a single-batten stretcher, or two battens, called a double-batten stretcher. Some side stretchers have unique names, such as "base stretcher" and "encircling legs".

- 黄花梨木炕桌（明）
  Yellow Rosewood *Kang* Table (Ming Dynasty, 1368-1644)

管脚枨：管脚枨是安装在椅子、凳子、桌子四条腿间下部的枨，因安装位置靠近足部，增加了家具腿部的牢固性，故名。有些管脚枨是在同一水平面上安装的，称为"四面平"；有些管脚枨为错开榫卯位置不在同一水平面上安装，一般是前面的横枨安装位置最低，两侧的横枨略高，后面的横枨最高，名曰"步步高"，有"步步高升"的吉祥寓意。

Base stretcher: stretcher installed in the lower part between the four legs of chairs, stools and tables. Because it is installed near the legs and enhances their firmness, it is known as *Guan Jiao Cheng* in Chinese, meaning stretcher that takes care of the legs. Some base stretchers are installed at the same level, called "straight form"; some are installed otherwise, arranged with the front one lowest, the side ones higher and the back one highest so that the joints do not overlap. This is called "stepped stretcher", an auspicious symbol of "stepping higher and higher".

- 铁力木官帽椅（明）
  Official Hat-shaped Ironwood Armchair
  (Ming Dynasty, 1368-1644)

罗锅枨：罗锅枨是一种中间部位向上凸起的曲形"横枨"，具有曲直的线条美，常与矮老与卡子花搭配。罗锅枨也具有实用功能，如安装在方桌上，相当于提高了横枨的位置，加大了桌子下部的空间，使用时更为舒适。

Humpbacked stretcher: a curved "side stretcher" with convex middle part. It displays the beauty of curves and straight lines and usually goes hand in hand with the pillar-shaped strut and decorative strut. The humpbacked stretcher is also very pragmatic. For instance, when installed in a square table, it elevates the position of side stretchers and increases the lower space, making the table comfortable to use.

- 黄花梨木方桌（明）
  Yellow Rosewood Square Table
  (Ming Dynasty, 1368-1644)

十字枨：十字枨是因安装位置不同而形成的。方桌、方凳等家具的横枨一般安装在相邻的腿之间，而十字枨则安装在对角位置的腿足之间，因两根横枨作"十"字形相交而得名。十字枨是稳定的三角形结构方式，因此比一般的横枨更加牢固。此外，六方形、八方形的家具也有采用类似方法装枨的，在六条或八条腿足间使用三根横枨或四根横枨，也是非常牢固的结构方式。

Crossed stretcher: stretcher formed by different installation positions. Side stretchers of square tables and square stools are commonly installed between adjoining legs while crossed stretchers are installed between diagonal legs and are so called for the cross-shaped intersection. Crossed stretcher is a stable triangular structural form and is therefore more firm than ordinary side stretchers. Besides, hexagonal and octangular furniture adopts the same method to install stretchers, using three or four side stretchers between six or eight legs, which is also firm in structure.

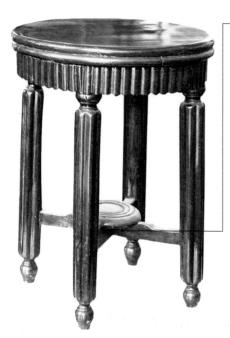

- 圆凳（近代）
  Round Stool (Modern Times, 1840-1919)

裹腿枨：裹腿枨是桌、椅、凳、榻、几等传统家具为仿竹藤家具的装饰风格而采用的一种特殊的横枨，看起来横枨要从外面裹住立腿。

Leg-encircling stretcher: a special side stretcher adopted for traditional furniture imitating the decoration style of cane-woven furniture, such as tables, chairs, stools, day beds and desks. It looks as if the legs are encircled by side stretchers.

- 藤面方凳（清）
  Square Stool with Cane-woven Surface (Qing Dynasty, 1644-1911)

霸王枨：霸王枨是一种不用横枨加固腿足的榫卯结构，主要用于方桌、方凳。制作造型清秀的桌子，嫌四条横枨碍事，但又要兼顾桌子牢固，就要采用"霸王枨"。霸王枨为S形，上端与桌面的穿带相接，用销钉固定，下端与腿足相接（位置在本来应放横枨处的里侧）。枨子下端的榫头为半个银锭形，腿足上的榫眼是下大上小。装配时，将霸王枨的榫头从腿足上榫眼插入，向上一拉，便勾挂住了，再用木楔将霸王枨固定住。

Giant's arm brace: a mortise and tenon joint for the reinforcement of legs that use no side stretcher, largely applied to square tables and square stools. In making tables with a simple molding, when four side stretchers are a hindrance but firmness is also needed, a giant's arm brace is adopted. It is S-shaped, with the upper part connected to the table surface with a penetrating transverse brace and fastened with wood or bamboo nails. The lower part is connected to table legs (in fact, it should have been installed inside the side stretcher). Tenons at the lower part of the stretcher are semi-dovetailed and mortises on the legs feature big bottoms and small tops. At the installation, insert the brace's tenons into the legs' mortises, pull upward to hook and plug tenon joints, then fasten the giant's arm brace with wooden wedges.

- **方桌（明）**
  Square Table (Ming Dynasty, 1368-1644)

横枨、罗锅枨、霸王枨、管脚枨、十字枨等。

腿式

腿式是指家具腿的式样，是家具的形体语言，对家具的造型影响较大，有方形直腿、圆柱直腿、扁圆腿、三弯如意腿、竹节腿等。还有的腿式在中部有束腰，有的还雕

Legs

Legs here refer to the leg patterns of furniture, the body language of furniture. They have great impact on furniture molding and mainly include square straight legs, cylinder straight legs, oblate legs, cabriole legs and bamboo joint legs. Some leg patterns feature a waist in the middle and some feature carved convex decorative motifs or beast heads.

- 螳螂腿

清代家具常用腿式之一，腿足上粗下细，呈S形，至脚头带弯外翻，形式柔媚而富有弹性，其形如细长的螳螂足，故名。

Mantis Leg

This is a common leg pattern of the Qing furniture, featuring a soft and flexible form and S-shaped legs with a thicker upper part that ends with outward-curving feet. The legs are slim and long like those of a mantis, hence the name "mantis leg".

- 一腿三牙

一腿三牙指桌子腿与牙子的结构方式。明代家具中有一种桌子，其四条腿中的任何一条都和三个牙子相接。三个牙子即两侧的两根长牙条和桌角的一块角牙，故名。

Three Spandrels to One Leg

This is a structural form of legs and spandrels. In the Ming furniture there is a type of table with four legs connected to three spandrels each, namely two apron-head spandrels on either side and a spandrel around the outer edges — hence the name "three spandrels to one leg".

- **三弯腿香几（清）**

  三弯腿是明清家具常用腿式之一。家具腿用料一般是圆或方，但有将脚柱在上段与下段过渡处向里挖成弯折的，这种腿足又大多有凸起的或外翻的脚头，故名。

  Cabriole-legged Incense Stand (Qing Dynasty, 1644-1911)

  Carbide legs are a leg pattern commonly employed in the Ming Dynasty (1368-1644) and Qing Dynasty (1644-1911). Although furniture legs are usually round or square, some are S-shaped legs ending in convex or outward-curving feet, which are the cabriole legs.

- **彭牙鼓腿圆形绣墩（清）**

  彭牙鼓腿是腿子和牙子都向外鼓出的做法。这种结构方式的优点是能以木材自身形状的变化来加强材料的质地、肌理的表现力。

  Embroidery Stool with Convex Apron and Bulging Leg Ending in a Horse-hoof Foot (Qing Dynasty, 1644-1911)

  Convex apron and bulging leg ending in a horse-hoof foot is a leg pattern that can depend on the transformation of pieces of wood to improve texture and enhance the expressiveness of grains.

- 圆形带托泥三弯腿香几（清）
Round Cabriole-legged Incense Stand with Continuous Floor Stretchers (Qing Dynasty, 1644-1911)

- 方茶几（明）
Square Tea Table (Ming Dynasty, 1368-1644)

刻凸出的花形或兽首。

托泥

托泥位于家具足部之下，一般是随着家具的底足位置随形制作，有圆形、条形或方框形。因其作用是将家具足部托起，使之不直接接触地面，故叫"托泥"。做工讲究的"托泥"下面还有小底足，俗称

Continuous Floor Stretcher

Continuous floor stretchers are located beneath furniture feet and shaped like circles, strips or frames, depending on the feet location. They mainly serve to prop up furniture feet to prevent them from touching the floor, hence the name "continuous floor stretcher". Under the

"龟足"。托泥使家具显得庄严厚重、精致考究。"托泥"和束腰一样,也是传统家具的造型手法之一。常见带托泥的家具有香几、案几、桌、坐墩、绣墩、扶手椅、画案等。

帽子

一些传统家具（如柜、屏风、架子床等）的顶部突出的装饰性部件被称作"帽子"。屏风帽子位居

exquisitely made floor stretchers are small feet, also known as "tortoise feet". Those stretchers make furniture magnificent and refined. Like the waist, they are also a molding of traditional furniture. Furniture with continuous floor stretchers include incense stands, recessed-leg tables, sitting stools, embroidery stools, armchairs and recessed-leg painting tables.

Cap

The protruding decorative member of some traditional furniture (cabinet, screen, canopy bed etc.) is referred to as a "cap". The cap of a screen is located right in the middle, with the

帽子
Cap

- 带帽子神龛（清）
  Capped Shrine(Qing Dynasty, 1644-1911)

屏风正中，一般中间稍高，两侧稍低，至两端又稍翘起，形如僧人所戴的帽子，故名"毗卢帽"，能把几扇屏风合拢在一起，起加固屏风的作用。此外，大型屏风帽上还有精美的浮雕纹，装饰性强，增添了屏风的气势。

矮老和卡子花

矮老是传统家具上的装饰构件，专指桌案、凳椅、床榻等家具

middle part of each section higher than the two sides and two upward turning cap ends, resembling a hat worn by monks if formed in a circle, hence its name "Vishnu Lou cap". It can join several screens together and serve as reinforcement. A large screen cap can have fine motifs in relief carving, which are quite decorative and add to the screen's imposing appearance.

Short Pillar-shaped Strut and Decorative Strut

The short pillar-shaped strut is a decorative piece of traditional furniture, exclusively referring to the tiny stretcher that serves as a prop between the apron and side stretcher of desks, chairs,

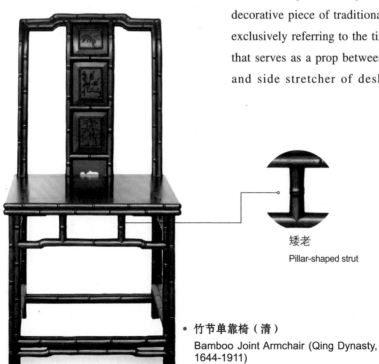

矮老
Pillar-shaped strut

• 竹节单靠椅（清）
Bamboo Joint Armchair (Qing Dynasty, 1644-1911)

• 卡子花
Decorative Strut

在牙条与横枨之间起支撑作用的小立枨，因其通常都不高，故名。矮老的外形要与家具风格保持一致，可单件使用，也可几件为一组。另外，在落地博古架与橱柜的腰间或底部，也有装配"矮老"的。

卡子花是一种图案化的"矮老"，常被雕刻成方胜、卷草、云头、玉璧、铜钱、花卉、双套环等形状，既起到对"矮老"的加固作用，又有较强的装饰作用，是传统家具重要的装饰手法之一。

搭脑

搭脑指高靠背椅子顶端的横挡，因正好位于人的后脑勺而得名。搭脑是高靠背的椅子才有的结构部件，在椅子造型和装饰方面起

and beds among others. It is so named because it is often not tall. The short pillar-shaped strut should be modeled according to the furniture style, used either individually or as a group. The short pillar-shaped strut is also installed at the waist or bottom part of floor antique shelves and cabinets.

The decorative strut is a short pillar-shaped strut with decorative motifs, commonly carved with designs such as overlapped squares, curling tendrils, cloud heads, jade discs, copper coins, flowers and double interlocking circles. It not only reinforces furniture as other short pillar-shaped struts do but is also highly decorative, making it an important element of decorative traditional furniture.

Top Rail

The top rail refers to the top horizontal rail of chairs with a high back. Since it is located right at the back side of the head, it is called *Da Nao* in Chinese (literal meaning: head's resting place). The top rail is only found in chairs with a high back and is so crucial to their molding and decoration that the chair's style is largely shaped by the top rail. There are three types of top rail: the first is a top

着很重要的作用，可以说，椅子的式样与搭脑的形状关系甚大。搭脑有三种类型：第一种是搭脑与椅子后立柱、扶手保持相同造型风格，如官帽椅、玫瑰椅等；第二种搭脑是中间加厚型，如太师椅的搭脑；第三种搭脑是花色型，一般多雕刻成吉祥物造型，如洋花椅等。

rail that maintains the same molding style as that of chair's back post and arms, such as the official hat-shaped armchair and rose armchair; the second is a top rail with a thickening middle part, like the top rail of *Taishi* (senior grand tutor) chairs; the third is a carved top rail, most often with carved mascots, such as *Yang Hua* (Chinese yellow azalea) armchairs.

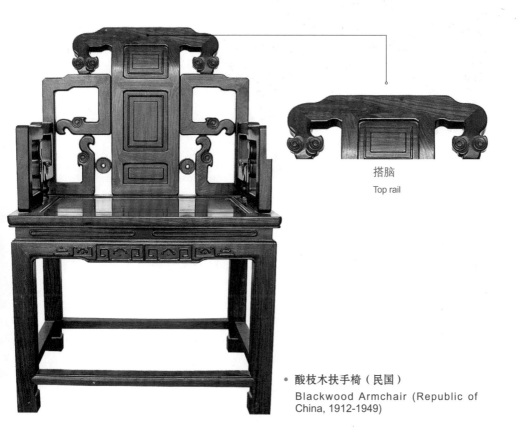

搭脑
Top rail

• 酸枝木扶手椅（民国）
Blackwood Armchair (Republic of China, 1912-1949)

搭脑
Top rail

• 搭脑扶手椅（清）
Armchair with Top Rails (Qing Dynasty, 1644-1911)

金属饰件

　　箱、柜、闷户橱等传统家具上有一些金属饰件，一般以铜制作，具有一定的实用功能和装饰效果，如面叶、吊牌、吊环、牛鼻环、套脚、合页、拍子、包角等。

Metal Ornaments

Traditional furniture, such as chests, cabinets and pantry cabinets, have metal ornaments usually made of copper that serve both practical and decorative functions. Such metal ornaments include plate (*Mianye*), hang tag (*Diaopai*),

lifting handle (*Diaohuan*), nose-ring handle (*Niubihuan*), foot cornerite (*Taojiao*), hinge (*Heye*), hasp (*Paizi*) and cornerite (*Baojiao*).

- 牛鼻环

带铜拉手的一种面叶，元宝式拉手就通过"牛鼻环"的结构挂在面叶上。

Nose-ring Handle (*Niubihuan*)

Niubihuan is a type of *Mianye* with copper handles. Its ingot-shaped handles are hung on the *Mianye* through a *Niubihuan* structure, i.e. a structure like the nose of cattle.

- 拍子

专门用于箱子上的铜质饰件，包括分为上下两半的面叶。拍子用转轴固定在上半个面叶上。拍子和上半个面叶安装在箱盖上，另半个面叶则固定在箱体上。箱子开启时，拍子起吊牌作用；箱子闭合时，拍子与下半个合页起扣吊作用，可以加锁。拍子上錾有花纹，做工精美。

Hasp (*Paizi*)

Paizi is a copper ornament used especially on chests, which includes the upper and lower plates (*Mianye*). The hasp (*Paizi*) is fixed on the upper half plate (*Mianye*) with a spindle, which as a whole is installed on the chest cover; while the other half plate is fixed on the chest body. When opening the chest, a *Paizi* serves as a hang tag (*Diaopai*); when the chest is closed, *Paizi*, together with the lower hinge, serves as a hasp, on which a lock can hang. Most *Paizi* are delicately furnished with decorative patterns.

- 合页

用于柜门上连接件，使门板便于开合。

Hinge (*Heye*)

Hinge is a connecting piece for opening and closing cabinet doors more easily.

- 面叶

面叶是传统家具上铜质饰件，多为片状的铜板，用扭头（外端有圆孔，可供穿锁用，其内端穿过面叶和柜门，将面叶固定在柜门上）和屈曲（用扁铜条对折而成，内端穿过面叶和柜门，将面叶固定在柜门上，其外端上的圆环孔可挂吊牌之类的拉手）固定在柜门上。

Plate (*Mianye*)

Mianye, usually slices of copper, are metal ornaments on traditional furniture fixed on cabinet doors by *Niutou* (with a round hole outside for hanging a lock and its inside passing through the door using *Mianye* for fixing it in place) and *Ququ* (a flat copper bar, the inside passing through the door using *Mianye* for fixing it in place, and the outside having a round hole for hanging metal pull and other handles).

- 吊牌

铜拉手之一种，式样很多，但均为片状的铜饰件，多与面叶配合使用。

Hang Tag (*Diaopai*)

Diaopai is one type of copper handle that comes in various styles, usually made of copper slices and used in conjunction with plate (*Mianye*).

精美的装饰工艺
Exquisite Decorative Crafts

　　中国传统家具不仅具有很强的实用性，而且装饰工艺十分精湛。这些工艺主要包括雕刻装饰、镶嵌装饰和漆饰等。正因为有了精良的装饰工艺，才使中国的传统家具成为珍贵的艺术品，备受世人追捧。

Traditional Chinese furniture is not only highly practical but also decorated exquisitely. The decorative craft includes carving, inlaying and painting. Thanks to such excellent decorative craft, traditional Chinese furniture pieces have become precious works of art and have been very popular for generations.

> 雕刻装饰

传统家具的雕刻装饰十分丰富，主要包括线刻装饰、浮雕装饰、透雕装饰、圆雕装饰、线脚装饰等。

浮雕

浮雕是在家具表面进行减地雕刻（将多余的木料去掉，使图案凸起来）而成的半立体形象，表现力十分丰富。

浮雕适合表现场面大、内容复杂的画面，如山水风景、楼台殿阁、街市等。通过浮雕底层到浮雕最高面的形象之间互相重叠、上下穿插的关系，使内容展现得深远和丰满。传统家具的浮雕装饰可根据浮雕的厚度划分为高浮雕、浅浮雕、中浮雕及深浮雕。

> Carving Decorations

Traditional furniture has rich carving decorations mainly including line carving, relief, openwork, three-dimensional carving and skintle.

Relief Decoration

Relief refers to carving on the surface of the furniture by clearing away redundant wood to make the pattern stand out from the surrounding background. The semi-three-dimensional pattern has strong expressive power.

Relief excels in presenting grand scenes with complex content, such as the landscape, attic and pavilion, and the marketplace. It can manifest rich and profound content through overlapping and interacting images from the top to the bottom level of the pattern. According to the thickness, relief decorations on traditional furniture consist of high relief,

云纹是明清家具上常用传统纹样,由龙纹和云纹组成,龙是主纹,云是辅纹,龙或驾云疾驰,或舞动于云间。

Cloud-dragon patterns are common on furniture of the Qing Dynasty (1644-1911). Its consist of cloud patterns and dragon patterns. Dragon patterns are major ones and cloud patterns, auxiliary, with dragons flying and dancing among the clouds.

- **紫檀木雕云龙纹宝座(清)**

  宝座是宫廷中供皇帝日常生活使用的坐具,陈设在皇帝和后妃寝宫的正殿明间,象征皇权至高无上。

  Rosewood Throne with Cloud-dragon Patterns (Qing Dynasty, 1644-1911)

  A throne is the seat of an emperor for daily use that is displayed in the main hall of the emperor's room and the rooms of his concubines. The throne is a symbol of the superiority of imperial power.

亮格：三面券口牙子和栏杆透雕拐子龙纹，华丽精致。

Shelf (*Liangge*): Delicate and gorgeous openwork grass (*Guaizi*) dragon patterns are carved on the arch-shaped apron and handrails.

浮雕卷草纹：牙板上浮雕卷草纹。

Rolling grass relief: There are rolling grass relief patterns on spandrel

合页：铜质，连接柜门和柜体。

Hinge (*Heye*): Copper hinge connects doors with the cabinet body.

- 雕花亮格柜（清）

清代亮格柜注重实用性和装饰性，设置多层亮格和多个抽屉，雕刻精美，有的镶嵌景泰蓝装饰片，隔板多用漆艺、漆画、瓷板画等。

Shelf-cabinet with Cnasty, (1644-1911)

The shelf-cabinet in the Qing Dynasty highlights both practical and decorative functions. It has multiple layers and several drawers with exquisite carved patterns. Some shelf-cabinets have cloisonné decorations and the layers are usually decorated with lacquer and porcelain paintings.

浮雕"四福捧寿"纹：左右门扇的上面各浮雕"四福（蝠）捧寿"纹，四只蝙蝠环绕团寿飞翔。"蝠"与"福"谐音，寓意福气、幸福。

"*Si Fu Peng Shou*" relief: Each of the two doors has a "*Si Fu Peng Shou*" relief pattern at the top — four bats flying around the Chinese character "*Shou*" (longevity). The Chinese character "*Fu*" is pronounced the same as "bat" and means good luck and happiness.

浮雕"幸福吉祥"纹：左右门扇的下面各浮雕"幸福（蝠）吉祥"纹，拐子纹中间悬挂飞翔的蝙蝠图案，寓意福在眼前，幸福吉祥。

"*Xing Fu Ji Xiang*" relief: There is a "*Xing Fu Ji Xiang*" relief pattern at the bottom of each of the two doors — a flying bat is surrounded by twisting patterns, meaning that good luck and happiness is right before your eyes.

另外，浮雕可以与圆雕结合使用，即用圆雕技法表现主要形象，以浮雕、线刻等技法表现其他次要形象并作为衬底。

圆雕

圆雕指不带背景、具有真实三度空间关系、适合从多角度观赏的雕

low relief, mid-relief and deep relief.

Moreover, relief can be combined with three-dimensional carving, with the latter presenting main characters and the former, together with line carving, presenting subordinate characters and serving as the background.

Three-dimensional Carving

Three-dimensional carving refers to the style of carving with three real dimensions and without background, suitable for

浮雕狮子戏球纹：狮为百兽之王，是权力与威严的象征。狮子戏球纹是中国传统的吉祥纹样，通常由气势威猛的雄狮构成，多用于家具、建筑细节的装饰。

Relief decoration of lions playing with balls: The lion is the king of animals, representing authority and dignity. The pattern of lions playing with balls is a traditional Chinese image to show good luck. It usually consists of muscular lions and is commonly used for decoration on furniture and buildings.

透雕脚踏：由两个半圆拼成，透雕几何形式的冰格纹，简洁空灵，脚踏下有四腿，腿足之间安有素面牙子。

Openwork footboard: The footboard is made up of two semi-circles, decorated with simple and lively openwork geometric ice patterns. Under the footboard are four legs and between the feet are plain spandrels.

圆雕三弯腿：六条腿道劲有力，上粗下细，中部雕刻龙纹，腿足外翻，足端刻虎爪抓珠。

Three-dimensional carved legs: The six legs of the table are strong and vigorous with thick tops and thin bottoms. Dragon patterns are in the middle. The feet are rolling outwards with patterns of tiger claws grasping pearls.

- 雕狮圆桌（清）
  Lion-pattern Round Table (Qing Dynasty, 1644-1911)

刻。传统家具的端头、柱头、腿足、底座等，多雕刻成人物、动物、植物形状，可以当做圆雕来看待。

还有一种半圆雕技法，常用来表现既有人物又有背景的纹样。其特点是主要形象用圆雕技法表现，次要形象和场景均用浮雕或线刻等技法来表现。

透雕

透雕主要用于家具上的牙板、围栏、环板、屏心、花板等部位，使家具展现出精工、通透、灵秀、华美的特色。

古代的透雕工艺是先将图案画在棉纸上，再将棉纸贴在木板上，然后在每组图案的空白处打一个孔，将钢锯丝穿入，往复拉动锼弓

appreciating from multiple angles. Traditionally, the top, stick tip, legs and feet and base of traditional furniture are carved into images of people, animals and plants that can be viewed as three-dimensional carvings.

Semi-three-dimensional carving is usually used in patterns with both people and a background. Major characters are displayed in three-dimensional carving, and minor ones and the background in relief decoration and line carving.

Openwork Decoration

Openwork decoration is mainly used on spandrel, rails, ring plates, screen centers and plates with patterns, making the furniture elegant and lively.

Ancient openwork technique initially involved drawing patterns on the tissue

- 圆雕"狮子滚绣球"（罗汉床局部）

  狮子在中国古代被视为"百兽之王"，外貌威严，是权力的象征。狮子滚绣球常用作喜庆的象征。

  Three-dimensional Carving A Lion Rolls a Silk Ball (part of the *Luohan* bed)

  The lion is regarded as the king of animals in ancient China. Its majestic appearance stands for authority. The pattern of A Lion Rolls a Silk Ball usually represents joy.

● 透雕"松鹤长春"纹（多宝格局部）

鹤在中国古代被视作仙禽，长颈、素羽、丹顶，传说其寿可至千年，因而被古人视为长寿鸟。

The Openwork Pattern of Cranes Playing Among Evergreen Pine Trees (Part of the *Duo Bao Ge*, Meaning Multi-grid)

The crane is regarded as an immortal animal in ancient China. It has a long neck, plain-colored feathers and a red crown. Legend says it can live for 1,000 years and was viewed as a long-living bird by the ancient Chinese.

子，沿图案的轮廓将空白处的木料"锼"走，因此又称"锼活"。锼好的半成品由专门的匠师进行细部精细的雕刻加工。

传统家具中，有些装饰采用透雕与多层次的深浮雕相结合的方式，具有较丰富的表现力。这是明清大型家具上常使用的一种雕刻技法。

and then sticking the tissue on a wood plate and drilling a hole in a blank space in the pattern. A hacksaw blade was put through the hole and pushed and pulled, cutting the blank out and highlighting the patterns. Such technique is called *"Sou"* in Chinese. The semi-finished patterns were then processed by specialized craftsmen working on the detail.

Some traditional furniture combines openwork decoration with multi-layer deep relief, thus giving it strong expressive power. This is commonly seen on large furniture from the Ming Dynasty (1368-1644) and the Qing Dynasty (1644-1911).

透雕百子图：百子图的典故源于周文王生百子的传说，是祥瑞之兆。画面常画众多小孩，寓意多子多孙，多福多寿。

One-hundred-children openwork pattern: The story behind the pattern originates from King Wen of Zhou Dynasty having 100 children, representing an auspicious sign. The pattern is usually full of children, meaning a lot of future generations, much luck and longevity.

- 楠木透雕百子图架子床（清）

  Phoebe Zhennan (*Nanmu*) Shelf Bed with One-hundred-children Openwork Pattern (Qing Dynasty, 1644-1911)

梅花纹：梅花寒冬开放，冰肌玉骨，代表着高尚的情操、坚忍不拔的精神品质。

The plum blossom pattern: The plum blossom is in full bloom in winter, pure and noble, representing noble sentiment and a tough spirit.

兰花纹：兰花是中国传统名花，以香气著称。兰叶终年常绿，兰花幽香清远，以气清、色清、神清、韵清的品性，被誉为"花中君子"，代表知交或有德行的人。

The orchid pattern: The orchid is a traditional Chinese flower well known for its fragrance. Its leaves are green all year long and its fragrance is delicate and far-reaching. The orchid is honored as the "gentlemen of flowers" for its fresh smell, color and charm, standing for bosom friends or people of good virtue.

竹子纹：竹四季常绿，临霜不凋，拔节向上，虚心有节，作为传统吉祥纹样，寓意子孙众多，品行高洁。

The bamboo pattern: Bamboo is evergreen even in cold winters. It grows upwards in a humble way and is used as a traditional pattern standing for good luck and a lot of future generations. Bamboo also represents noble behavior.

菊花纹：菊花凌霜盛开，花色素雅，被赋予了吉祥、长寿的含义，象征着名士的孤高亮节与高雅的友情，也是中华民族不屈不挠精神的体现。

The chrysanthemum pattern: The chrysanthemum has simple but elegant colors and blossoms in the autumn frost. It is regarded as the symbol of auspiciousness and longevity, representing exemplary conduct and the true friendship of celebrated people who have no official positions. It also embodies the Chinese spirit of perseverance.

- **透雕"四君子"纹（座屏局部）**

  梅、兰、竹、菊是中国古人喜爱的植物，并称为花中"四君子"。

  "The Four Gentlemen" Openwork Decoration (Part of the Screen with Pedestal)

  "The Four Gentlemen" refers to four favorite plants in ancient China — the plum blossom, the orchid, the bamboo and the chrysanthemum.

八仙纹是指汉钟离、吕洞宾、铁拐李、曹国舅、蓝采和、张果老、韩湘子、何仙姑八位仙人组成的纹样,他们是民间流传的八位神仙。八仙纹常作为寓意吉祥的图案广泛用于家具装饰。

The pattern of Eight Immortals refers to a picture consisting of Han Zhongli, Lü Dongbin, Tie Guaili, Cao Guojiu, Lan Caihe, Zhang Guolao, Han Xiangzi and He Xiangu, the eight immortals popular in folklore. The pattern is widely used on furniture as a symbol of auspiciousness.

- 透雕八仙供桌(清)
  Eight Immortals Table with Openwork Patterns (Qing Dynasty, 1644-1911)

线脚装饰

传统家具的大边、抹头或腿足的截断面并不都是四方（含长方形）形或圆（含椭圆）形，大多数呈不同凹凸起伏的形状，看上去很舒服，手感也很好，这就是传统家具上的"线脚"。例如，像竹子一样呈圆弧形凸起的线脚装饰，叫"竹片浑"。

Skintle Decoration

The bedside, window stool and cross-sections of legs and feet of traditional furniture are not always square or rectangular or round or oval. Instead, most of them are rather uneven, comfortable for both viewing and touching. This is called the "skintle" of furniture. For instance, *"Zhu Pian Hun"* is a type of such decoration, as round as bamboos.

竹片浑：指像竹片那样圆弧形的浑面。竹片浑流行于清代中晚期，朴实优美的竹节造型使家具造型自然，表现出独特的艺术效果。

Zhu Pian Hun: Cross-sections as round as bamboos. It was popular during the middle and late Qing Dynasty. The simple and elegant bamboo-shaped decoration makes furniture natural in formation and unique in artistic effect.

- 棋桌（清）
  Chess Table (Qing Dynasty, 1644-1911)

拦水线：酒桌的桌面四周常设有一道阳线，叫"拦水线"，以免在酒肴倾洒时弄湿衣服。

Water-barring skintle: The high relief molding around the wine table is called *Lan Shui Xian*. It prevents clothes from getting wet from spilt wine.

• 酒桌（明）
Wine Table (Ming Dynasty, 1368-1644)

传统家具之所以要采用线脚装饰，是因为其主要靠烫蜡和木质本色来起装饰作用，一般都有光反射弱、层次感差的缺点。凹凸不平的线脚装饰，会对光有不同的反射效果，使家具轮廓线产生变化，出现丰富的层次感。

传统家具常用的线脚有许多种，包括阳线、凹线、线香线、皮带线、拦水线、弄洞线、芝麻梗、竹片浑等。

If the decoration on traditional furniture relies only on melted wax and the natural color of wood, the furniture may lack light reflections and the effect of multiple tiers. With uneven skintle decorations, the furniture reflects lights in different ways and looks more varied and textured.

Common skintle decorations on traditional furniture include skintle in high relief (*Yang Xian*), skintle in intaglio (*Ao Xian*), fragrant skintle (*Xian Xiang Xian*), leather belt skintle (*Pi Dai Xian*), water-barring skintle (*Lan Shui Xian*), holing skintle (*Nong Dong Xian*), sesame rice (*Zhi Ma Geng*) and bamboo skintle (*Zhu Pian Hun*).

> 镶嵌装饰

镶嵌装饰即是将不同的材料制成饰物镶嵌在家具上，起到装饰的作用。因所用材料不同而有不同的名称，常见的有木嵌、螺钿嵌、牛骨嵌、银丝嵌、珐琅嵌、嵌大理石、嵌瓷板画、嵌象牙雕刻等。

> Inlay Decorations

Inlay decoration refers to inlaying ornaments made of different materials on furniture for the purpose of decoration. Common inlay decorations include wood, mother-of-pearl, ox bone, silver thread, enamel, marble, painted porcelain and carved ivory.

- **束腰珐琅面心方凳（清）**

  珐琅嵌即用珐琅工艺制成平板状的各种饰片，然后镶嵌在家具相应的表面，使家具产生一种豪华的风格。这种技术多见于清代家具。

  Bundled Square Stool with Enamel Inlaid Patterns (Qing Dynasty, 1644-1911)

  Enamel inlay refers to inlaying various decorative pieces made of enamel into the surface of furniture, making the furniture luxurious. Such a technique is commonly seen in Qing furniture.

- **掐丝珐琅挂屏（清）**
  掐丝珐琅一般在金、铜胎上以金丝或铜丝掐出图案，然后填上各种颜色的珐琅，经过焙烧、研磨、镀金等多道工序制成。

  Cloisonné Enamel Screen (Qing Dynasty, 1644-1911)
  Cloisonné enamel refers to filigreeing patterns on gold and copper models and filling them with colorful enamel. It also consists of many procedures including roasting, polishing and gilding.

螺钿"寿"字：清代螺钿镶嵌十分细密，还采用了金、银片。桌面中间螺钿"寿"字共120字，沿嵌螺钿"卍"字锦纹地，寓意万寿无疆。

Mother-of-pearl inlaid pattern of "*Shou*": Mother-of-pearl inlay of the Qing Dynasty (1644-1911) is very detailed and uses gold and silver pieces. There are 120 "*Shou*" characters at the center of the table with patterns of 卍 around the edge, representing longevity.

- 嵌螺钿炕桌（清）

  螺钿嵌是用较厚的贝壳制成人物、花草、动物等纹件，然后镶嵌在传统家具上，打磨平整，表面上不施漆。清代中期以后，采用这种装饰技法的家具很多。此桌面螺钿为"寿"字，侧沿螺钿连"卍"字纹，面下束腰，嵌团寿及长寿字纹，牙条及直腿螺钿蝙蝠、寿桃、团寿及方寿纹，寓意福寿双全。

  Kang Table with Mother-of-pearl Inlay (Qing Dynasty, 1644-1911)

  Mother-of-pearl inlay refers to inlaying patterns of people, plants and animals that are made of thick shells into furniture and then polishing the surface of the patterns without painting. Such a decoration technique is commonly seen after the middle of the Qing Dynasty. On this *Kang* table, patterns of "*Shou*" (a Chinese character, meaning longevity) are on the table surface, and patterns of 卍 are around the side edge. The surface is bundled beneath and inlaid with round and long patterns of "*Shou*", and the *Ya Tiao* and legs are inlaid with patterns of bat, peach and round and long "*Shou*", representing both good luck and longevity.

铁力木大理石面心炕桌（清）

嵌大理石即在家具上嵌大理石作为面板。大理石有美丽的变化无穷的纹理，特别是那种酷似某种自然形象的纹理，更令人叹为观止。传统家具一般选用上品大理石，以白如玉、黑如墨者为贵；微白带青、微黑带灰者为下品。除大理石外，还有用如朝霞一般红润的红色玛瑙石、碎花藕粉色的云石、花纹如玛瑙的土玛瑙石等进行镶嵌的家具。

Ironwood-marble *Kang* Table (Qing Dynasty, 1644-1911)

Marble inlay refers to inlaying marble into furniture to serve as the surface. Marble has beautiful and varied patterns and that give the remarkable impression of representing natural images. Traditional furniture usually uses high-quality marble, in particular, marble as white as jade or as black as ink. Marble with cyan or gray flaws are considered low quality. Some furniture is also decorated with agate as red as morning rosy clouds, pink alabaster with fragmented flower patterns and turbid enamel with patterns like those of pure enamel.

嵌青花瓷画小座屏（清）

嵌瓷板画即在家具上嵌有彩绘纹样的瓷板作为面板。彩绘瓷是高档艺术品，用来装饰的家具也是十分名贵的。

Blue-and-white Porcelain Inlay Stand with Pedestal (Qing Dynasty, 1644-1911)

Porcelain painting inlay refers to inlaying porcelain plate with colored paintings into the surface of the furniture. Colored porcelain is a high-range work of art, so furniture with such decorations is very precious.

- 嵌银丝座屏（清）

银丝嵌即是在家具上装饰银丝图案作为面板。其工艺方法是先将白银加工成很细的银丝，并设计出适合家具各个部位的二方连续的纹样，然后将画有纹样的绵纸贴在家具的表面，干后，选用与纹样同形状的薄口小刻刀，依纹样凿刻出浅槽，每凿刻数刀，便将银丝压嵌入槽内。待全部银丝嵌完后，用木槌轻轻敲实至平，再经打磨后，便可上蜡或擦漆。此座屏用银丝装饰了唐代画家吴道子的《八十七神仙图》，描绘的是87位神仙列队前往朝拜元始天尊的情景。他们行走在亭台曲桥上，其间有流水行云等点缀，恍若仙境，空中似乎有仙乐飘荡。人物刻画生动传神，飘逸洒脱。

Silver Inlaid Stand with Pedestal (Qing Dynasty, 1644-1911)

Silver inlay refers to inlaying silver patterns on the surface of furniture. The technique is to process silver into thread and design a double repeating pattern suitable for every part of the furniture. Stick the tissue with patterns on the furniture surface and dry it. Choose a small thin-blade engraver with the same pattern as patterns on the tissue and carve the patterns out. While carving, push the silver thread into the carving slot and gently tap the thread flat in the end with a wooden hammer. Wax or paint can be added after polishing. This stand with pedestal has a silver inlaid picture of "Eighty-seven Immortals" by Wu Daozi (680-759), a painter of the Tang Dynasty. In the picture, 87 immortals are lining up to visit the original celestial being. They are walking either across temples and pavilions or on bridges. A running stream and flying clouds are around them as decoration. The scene is like a fairyland and there seem to be magic melodies waving in the air. The characters in the picture are lively and graceful.

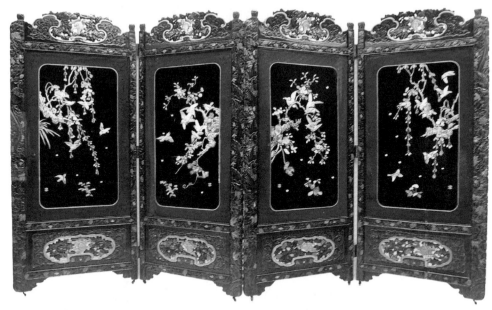

- 百宝嵌花鸟纹曲屏（清）

百宝嵌是在一件家具上嵌入多种名贵材料，如用玉、石、牙、角、玛瑙、琥珀等制成的装饰物。明清大漆屏风多采用"百宝嵌"工艺，即用玉、石、牙、角、玛瑙、琥珀等多种名贵材料雕成山水、人物、树木、楼台、花卉、翎毛等，嵌于漆器家具之上，非常美丽。

Folding Screen with Treasure Inlaid Pattern of Flowers and Birds (Qing Dynasty, 1644-1911)

Treasure inlay refers to inlaying several precious materials, including jade, stone, ivory, antler, agate and amber, into furniture. Most painted screens in the Ming Dynasty (1368-1644) and Qing Dynasty (1644-1911) adopted this technique, i.e. carving the materials mentioned above into patterns featuring mountains, rivers, people, trees, towers, flowers and feathers and inlaying them into painted furniture. The effect is impressive.

> 漆饰

漆饰是指用漆装饰传统家具。明清时期，漆饰工艺发展到顶峰，手法丰富多变，包括一色漆、罩漆、描漆、描金、堆漆、填漆、雕漆、刻漆等。

> Painting Decorations

Painting decoration involves embellishing furniture with lacquers. The technique reached its peak during the Ming Dynasty (1368-1644) and Qing Dynasty (1644-1911) and consisted of various styles including one-color painting, transparent painting, painting depicting, gold depicting, overlapping paining, filling painting, engraving painting and cutting painting.

- **剔红雕九龙纹宝座（清）**
  剔红，是在器物的胎型上涂上近百道朱色大漆，待干后雕刻出浮雕的纹样。
  Tihong Throne with Nine-dragon Patterns (Qing Dynasty, 1644-1911)
  Tihong refers to painting nearly 100 lines of scarlet oil paint on the surface of the furniture and then carving patterns in relief on the paint.

描金蟠螭纹：螭是中国古代传说中的一种神兽，头上无角，有四只脚和一条长卷尾。蟠螭纹是盘曲而伏的螭纹，构图呈半圆形或近圆形。

Panchi patterns in gold depicting: Panchi refers to a legendary beast in ancient Chinese folklore. It is hornless, has four feet and a long curly tail. Panchi patterns show the beast prone. The layout is usually semi-circular or circular.

描金"太平景象"图案："瓶"与"平"同音，象是瑞兽。图案为象驮宝瓶或象鼻卷瓶，瓶中插有花卉，有安宁和平、民康物阜的吉祥寓意。

"*Tai Ping Jing Xiang*" pattern in gold depicting: In Chinese, "bottle" sounds the same as "*Ping*" (safety), and "*Xiang*" is a beast symbolizing auspiciousness. In the pattern, a *Xiang* is depicted carrying a precious bottle on the back or rolling it with its nose. There are flowers in the bottle, representing peace, safety, health and wealth.

- **楠木描金椅（清）**

  描金即在漆木家具表面用金色描绘花纹的漆饰技法，包括贴金和画金。

  Phoebe Zhennan (*Nanmu*) Gold-Depicting Chair (Qing Dynasty, 1644-1911)

  Gold depicting refers to drawing patterns using gold color on the surface of a painted piece of furniture. It includes gilding and drawing pictures with gold.

- 黄花梨木雕龙纹刻漆山水人物曲屏（清）

  刻漆一般用来装饰大面积的漆面，即在漆面上按图案轮廓剔去漆灰，然后在花纹内填入漆色或油色，呈现出图画效果。此曲屏的屏心描绘了十八学士，腰板描绘的是西湖十景。

  Yellow Rosewood Folding Screen with in Cutting-painting Patterns Depicting Mountains, Rivers and People (Qing Dynasty, 1644-1911)

  Cutting painting is usually used to decorate large paintings. The process is to carve the patterns out on the painting's surface and fill them with oil paint. This screen demonstrates the Eighteen Scholars in the middle and the top 10 tourist spots around the West Lake in its waist part.

• 现代家庭中的传统家具摆设
Modern Pieces of Traditional Furniture

传统家具的常用名贵木材
Precious wood commonly used in traditional furniture

木材名称 Name	木材图样 Pattern	木材特点 Features	用途 Functions
楠木 Phoebe Zhennan (*Nanmu*)		木质优良，纹理美观，结构致密，质地温润柔和，切面有光泽，偶尔还会散发出阵阵的幽香，具有很强的防潮防腐蚀性。 High quality, fine texture, compact structure, mild character, glossy cross-sections, occasionally fragrant, highly moisture-proof and non-corrosive.	楠木是造船、建筑、家具等的珍贵用材。楠木的木性温和，冬天触之不凉，常用来制作桌案、罗汉床、柜子和书架，也可用来装饰柜门或制作文房用具。 Precious material for ships, buildings and furniture. Mild in nature and warm to the touch in winter. Usually used for tables, *Luohan* beds, cabinets and book shelves, decorating cabinet doors and making study tools.
紫檀木 Rosewood (Red Sandalwood)		质地细密，木材的分量重，入水即沉；心材新切面颜色橘红色或鲜红色，历久后转为紫色或紫黑色，常带有浅色和紫黑色条纹。木材纹理交错，局部卷曲，呈绞丝状，导管隐约可见，如蟹行泥中的眼点，管孔内细密弯曲极像牛毛，也称之为"牛毛纹"。 Compact structure, sinks in water due to high density, new heartwood cross-sections are orange or scarlet and purple or dark purple after long with light-color and purple-black strips. Overlapping and sometimes rolling texture, indistinct thin pipes in the wood are like crab's eyes. The dense and tiny curves are like ox hairs, thus called "ox hair" patterns.	紫檀木在明清时期备受皇家推崇，大量用于家具制作、工艺品雕刻等。由于紫檀木的生长周期较长，数量稀少，因而存世的紫檀家具和工艺品均被视为稀世珍品，传世收藏。 Highly praised by the imperial class during the Ming Dynasty (1368-1644) and Qing Dynasty (1644-1911). Extensively used in furniture and carved works of art. Due to rosewood's long growth cycle and rarity, the existing pieces are regarded highly precious and passed down by generations.

黄花梨木 Yellow Rosewood		心材新切面呈紫红色或深红褐色，生长轮比较明显，纹理斜或交错，有奇趣可观的"鬼脸纹"。新切面的气味辛辣浓郁，久则微香。 New heartwood cross-sections are purplish red or dark crimson with clear growth rings, overlapping texture with bizarre "ghost-face" patterns. With strong fragrance from new cross-sections and light fragrance later.	黄花梨木木质细腻，自然美观，香气持久，工艺性能十分优越，缩胀率小，不易变形，手感温润，坚固耐腐，是专做贵重家具和雕刻工艺品的上等材料。明清的黄花梨家具被视作世界家具艺术中的珍品。 Exquisite texture, natural and beautiful, long-lasting fragrance, easy to process, low expansion and reduction ratios, unlikely to become deformed, mild to the touch, non-corrosive. High-range material for precious furniture and carved works of art. Yellow rosewood of the Ming Dynasty (1368-1644) and Qing Dynasty (1644-1911) is highly cherished in the global furniture industry.
鸡翅木 Wenge (Chicken-wing Wood) (Jambire)		木材心材的切面上有鸡翅（"V"字形）花纹，因其纹理酷似鸡翅羽毛而得名，纹理交错、清晰，颜色突兀，有微香气，生长年轮不明显。 Also called wenge for its chicken wing-shaped (V-shaped) patterns on heartwood cross-sections. Overlapping and clear texture, sharp color, lightly fragrant, indistinct growth rings.	鸡翅木非常适合做文房家具和茶具，因为鸡翅木在热水的刺激下，会挥发出一种很自然的香气，这种香气有提神的效果。 Very suitable for furniture in study rooms and tea sets for its natural and refreshing fragrance when heated.
酸枝木 Sections		坚硬耐用，心材一般为浅红褐色、红褐色、紫红色、橙色、紫褐色至黑褐色，材色不均匀，随着时间增长和空气氧化，材色逐渐变成深红色、黑红色。有光滑细密纹理，锯开木材时会产生刺鼻的酸味。 Hard and durable. Light red brown, red brown, purplish red, orange, puce and dark	酸枝木多用来仿制明代黄花梨家具。由于酸枝木的产量大，用其制作家具时可以选取最精美的部分，因此酸枝木家具也是品质很高的上等家具，逐渐成为紫檀木、花梨木等上等木材日见匮乏之后的替代品。 Usually used to copy yellow rosewood furniture of the Ming Dynasty (1368-1644). Due to large production, the best parts of

		red and black red over time through oxidation. With smooth and exquisite texture. Acrid smell when sawn.	blackwood can be sorted out to make furniture. Therefore, furniture made of blackwood is also very precious, gradually substituting furniture made from rosewood and yellow rosewood.
榉木 Bowed String Instruments		木质重且坚固、抗冲击、有光泽，强度大，耐摩擦、耐水浸；纹理直且清晰，质地均匀，常带有美丽的大花纹，以塔形纹最佳，俗称"宝塔纹"；色调柔和流畅，边材呈黄褐色，心材呈浅栗褐色带黄。 High density, hard, shock-resistant, glossy, scratch-and water-resistant, clear texture, well-distributed structure, large and with beautiful patterns (especially tower-shaped ones called "pagoda patterns"), with gentle and smooth color tone, yellowish brown sapwood and light brown heartwood.	榉木是中国明清时期民间家具最主要的用材，多用作家具、门、地板、室内细木工制品、镶嵌面板、刷子柄、车制品和食品容器等。活树多用作庭园美化和绿荫树。 The most commonly used wood among households during the Ming Dynasty (1368-1644) and Qing Dynasty (1644-1911), mainly for making furniture, doors, floors, indoor cabinetwork, panels, brush handles, car products and food containers, also for embellishing the environment and providing shade.
核桃木 Plucked String Instruments		核桃木的边材呈黄褐色至栗褐色，心材呈红褐色至栗褐色，有时显现紫色；间杂有深色条纹，有时带美丽的斑点或条纹；时间久置呈现深棕色，生长轮明显；木质管孔中含有深色树胶，有油脂；木质坚硬，具有明亮的光泽。很多核桃树都有树瘤，带有树瘤的核桃木具有特殊的工艺和装饰价值。 Yellow brown to chestnut brown sapwood, red brown to chestnut brown heartwood, sometimes purple, mixed dark stripes, sometimes beautiful spots or stripes, dark brown after long storage, clear growth	核桃木经过较为漫长的干燥程序后，不易变形或开裂，胶粘性质也很好，很适合开榫和雕琢。因此，核桃木一般用来制作高档家具、建筑结构、雕刻品、门、地板、镶板等。核桃木的树根和瘿瘤等常有特殊花纹，锯为薄木后是珍贵的装饰材料。 Hard to deform or crack after long drying process, good adhesiveness, suitable for tenoning and carving. Usually used to make high-range furniture, buildings, sculptures, doors, floors and panels. Unique pattern on the root and cecidium that make precious decorations after being sawn into thin slices.

rings, dark gum and fat in wood pipes, hard and glossy. Burls are common, adding unique artistic and decorative value to walnut.

| 榆木 Elm |  | 边材呈暗黄色，心材则呈紫红色。纹理通直、花纹明晰、棕眼显著。木质弹性好，坚韧、耐腐、耐湿。易于加工，一般透雕浮雕均能适应，刨面光滑，弦面花纹美丽，有"鸡翅木"的花纹。

Dark yellow sapwood and purplish red heartwood, straight and clear patterns, clear spots, good elasticity, corrosion-and moisture-resistant, easy to process, suitable for openwork and relief carving, smooth cross-sections, beautiful patterns on vertical-sections, e.g. "chicken wing" patterns. | 榆木家具的制作年代大约从明早期至清晚期。早期榆木家具大多用来做供奉家具，比如寺庙、家祠等处的供桌、供案。但由于榆木木质坚硬，制作起来比较费时费工，因此清晚期后便很少用来制作家具。此外，榆木经烘干、整形、雕磨髹漆后，可制作成精美的雕漆工艺品。

Elm furniture was popular during the early Ming Dynasty (1368-1644) to the late Qing Dynasty (1644-1911). Early elm furniture was used in temples and mourning halls. Later, it was seldom used to make furniture for its hard character makes it time-consuming to process. It can be made into fine carved lacquerware after drying, reshaping, polishing and painting. |

| 铁力木 Ironwood |  | 质坚而沉重，心材淡红色，结构均匀，纹理致密，具有强度大、耐磨损、抗腐蚀、防虫蛀的优点。

Hard, high density, light red heartwood, well-distributed structure, compact texture, wear-and corrosion-resistant, not susceptible to woodworm. | 铁力木用途很广，是高级家具、特种雕刻、抗冲击器具、珍贵镶嵌和高级乐器的理想用材，比如造船、车辆、建筑等。

Widely used, ideal material for making high-range furniture, special carving, shock-resistant tools, precious inlays and classic music instruments, also in ships, cars and buildings. |

经典家具式样鉴赏
Classic Furniture Appreciation

中国自出现家具以来，到明清时期进入了黄金时代，家具的样式在使用及生产加工的过程中日趋精美，逐渐形成了完备的体系。传统家具的经典式样主要有床榻类、椅凳类、桌案类、箱柜类、屏风类、支架类等。

The Ming Dynasty (1368–1644) and Qing Dynasty (1644–1911) were the golden period of furniture in China. Furniture styles had become more plentiful over time, forming a complete catalog. Classic furniture types include beds, couches, chairs, stools, tables, desks, cabinets, chests, screens and stands.

> 床榻类

床和榻都是供人睡卧休息的家具，人们习惯上将其并称。汉代以后，床专指卧具，榻则成为供官僚贵族和文人雅士会客休息所用的坐具。

明清时期的床，形制高大，宛如一间雕梁画栋的房子，强调密闭性。明清时期的榻，都是陈设在厅堂中的坐具。

明代的床一般较宽大，能睡双人，摆放在居室中的暗间，有架子床、宝座床（龙凤床）、拔步床等式样。明代的榻大多仅容一人坐卧，有罗汉床、宝座、贵妃榻、床柜等形式，一般陈设在正房明间，供主人休息和接待客人之用，其功能相当于现在的双人沙发。清代床榻在康熙以前保

> Beds and Couches

Beds and couches are furniture for resting. People often mention them together. After the Han Dynasty (206 B.C.-220 A.D.), beds referred especially to the furniture people lie on and couches referred especially to the seats that bureaucrats, nobles and scholars rested upon and at which they met guests.

Beds of the Ming Dynasty (1368-1644) and Qing Dynasty (1644-1911) were large and tall with carved beams and painted pillars, emphasizing integrity. Couches of the same period were displayed in the hall as seats.

Beds of the Ming Dynasty (1368-1644) were usually broad enough for two people to sleep in and placed in the inner room of the hall. The types of bed included shelf beds, throne beds (dragon-phoenix beds), *Ba Bu* beds and

床屉：架子床的床屉一般分为两层，下层为棕屉，上层为藤席，多见于中国南方地区。北方多用木屉，可防寒保温。

Bed drawers: Bed drawers of shelf beds usually have two layers: the upper one is a vine drawer and the lower one is a palm drawer. They are common in southern China. Wood drawers are common in northern China to keep warmth.

床座：由纵、横的方木制做成的长方形框。明清时期床座前面的前沿边缘多起阳凸线装饰，下有束腰，十分讲究。

Bed base: A bed base is a rectangle frame made of vertical and horizontal wood. Bed bases of the Ming Dynasty (1368-1644) and Qing Dynasty (1644-1911) had very prescribed decorations, with raised decorations along the bed front and waist structure beneath the bed surface.

立柱：架子床四角设有立柱。明清时期，还有在前面两柱间增设两柱，为六柱的。

Pillars: A shelf bed has four pillars at its corners. Two pillars were added to the bed front during the Ming Dynasty (1368-1644) and Qing Dynasty (1644-1911), taking the number of pillars to six.

围栏：明清时期床围栏多是用细木攒接成各种花纹的围板，非常精致，疏朗隽秀。

Rails: Bed rails of the Ming Dynasty (1368-1644) and Qing Dynasty (1644-1911) were made of thin wood with various exquisite and elegant patterns.

- **黄花梨木六柱式架子床（明）**

  架子床是有柱子承托床顶的双人床的统称，有多种形制，便于悬挂蚊帐、锦帐等。

  Yellow Rosewood Shelf Bed with Six Pillars (Ming Dynasty, 1368-1644)

  The shelf bed is a double bed with pillars supporting a roof. It came in various styles and was convenient for hanging mosquito and brocade nets.

- 榆木凉榻（清）
  四足榻的榻屉采用多根有间隙排列的木条或竹条做成，极为平整，四周不起沿，通风透气。

  Elm Summer Couch (Qing Dynasty, 1644-1911)
  The four-leg couch drawer is made of tiered wood or bamboo strips. It has no prominent edges and is ventilative.

留着明代的特点。到了乾隆时期，形成了用材厚重、装饰华丽的风格，家具制作力求繁缛精致，不惜耗费工时和木材。

so forth. Couches of the same period were only big enough for one people to sit on. They included the *Luohan* beds, throne seats, *Guifei* (concubine) couches, bed chest etc. They were usually placed in the central room of the hall for the host to rest upon and receive guests, serving as today's double couches. Beds and couches of the Qing Dynasty (1644-1911) kept the features of those of the Ming Dynasty (1368-1644). During the reign of Emperor Qianlong (1736-1795) they highlighted the luxurious use of materials and splendid decoration. Furniture was made through countless processes and detailed decorations regardless of the time, energy and materials.

踏步：长出床沿 1 米左右的平台，在床前形成一个小廊。

Footboard: A one-meter-long platform in front of the bed that forms a small corridor.

床顶：清代拔步床床顶安盖，中间镶接吉祥图案，雕饰繁缛。

Bed top: *Ba Bu* bed of the Qing Dynasty (1644-1911) had a cover on the top that was inlaid with auspicious patterns and other complicated decorations.

底座：是一个木制平台，床置于其上。

Base: A wooden platform beneath the bed.

- **拔步床（清）**

  拔步床形制庞大，从外观看好像一个木屋。床前有廊，有相对独立的活动范围，廊庑两侧可以放置桌凳、便桶、灯盏等小型家具。人跨步入廊犹如跨入室内，地上铺板，床置于地板之上。床架的作用是为了便于挂帐子，以免蚊虫叮咬。

  ### *Ba Bu* Bed (Qing Dynasty, 1644-1911)

  Ba Bu beds are very large and look like a wooden house from the outside. The bed front has a corridor as independent space for activities. A table, stool, chamber pot, lamp and other small items can be placed along both sides of the corridor. Stepping into the corridor, one feels like entering a room. The floor is paved and the bed is on the floor. The bed shelf is for hanging a mosquito net.

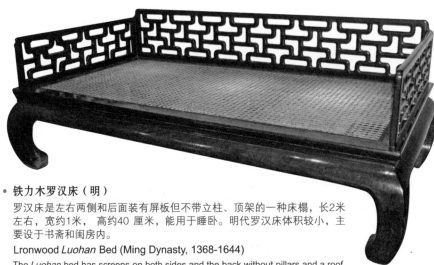

- 铁力木罗汉床（明）

罗汉床是左右两侧和后面装有屏板但不带立柱、顶架的一种床榻，长2米左右，宽约1米，高约40厘米，能用于睡卧。明代罗汉床体积较小，主要设于书斋和闺房内。

Lronwood *Luohan* Bed (Ming Dynasty, 1368-1644)

The *Luohan* bed has screens on both sides and the back without pillars and a roof. It is about two meters long, one meter wide and forty centimeters high. It was used for lying on and sleeping. *Luohan* beds of the Ming Dynasty (1368-1644) were usually small and placed in the study rooms and ladies' bedrooms.

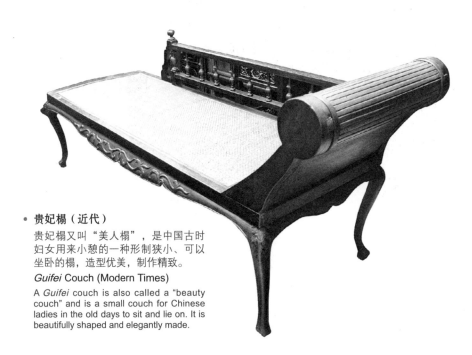

- 贵妃榻（近代）

贵妃榻又叫"美人榻"，是中国古时妇女用来小憩的一种形制狭小、可以坐卧的榻，造型优美，制作精致。

Guifei Couch (Modern Times)

A *Guifei* couch is also called a "beauty couch" and is a small couch for Chinese ladies in the old days to sit and lie on. It is beautifully shaped and elegantly made.

● 现代家庭中的传统家具摆设
Traditional Furniture Layout in Modern Houses

> 椅凳类

椅凳类家具包括椅子、凳子、坐墩、交杌等坐具,唐代中期以后正式进入上流社会,在宫廷宴饮、家居生活及行军战场中应用。

椅是传统家具中最具典型性的有靠背的高型坐具,最早出现在汉代。从明代起,椅的式样空前多了起来,而且把精巧、实用的传统美学观念和人体结构相结合,创造了简约的风格。清代的椅又有所变化,主要是增加了雕刻装饰,变肃穆为流畅,化简素为雍容,并加大了椅的尺度。

凳是汉代时出现的无靠背坐具。明代的凳有方圆两类,其中以方凳的种类最多。清代的凳以束腰为主,方凳、圆凳的尺寸也

> Chairs and Stools

In the middle Tang Dynasty (618-907), sitting furniture such as chairs, stools, sitting blocks and *Jiaowu* were officially used in the upper class society, such as imperial banquets, households and battle fields.

The chair is the most typical traditional sitting furniture with a back. It originated during the Han Dynasty (206 B.C.-220 A.D.). From the Ming Dynasty (1368-1644), there were more styles of chair and craftsmen started to combine traditional aesthetic concepts with human body structure, creating a very simple style. Chairs of the Qing Dynasty (1644-1911) underwent more changes, mainly an increase in carved decorations. Instead of solemnity, chairs became more stylistic, smoothly shaped and larger.

Stools or sitting furniture without backs emerged during the Han Dynasty (206 B.C.-220 A.D.). Stools of the Ming

较明代小些，式样清秀，宜在小巧精致的房间里摆放。

Dynasty (1368-1644) included two types, round and square. There were more styles for square stools. During the Qing Dynasty (1644-1911), stools were usually bundled beneath and smaller than those of the Ming Dynasty (1368-1644). They were very pretty and suitable for small rooms.

靠背上的透雕麒麟纹：麒麟是中国古代传说中的一种瑞兽，麒麟纹寓意吉祥，事业有成。

Kylin openwork decoration on the back: The kylin is an auspicious beast in ancient Chinese folklore and kylin patterns represent good luck, auspiciousness and business success.

椅面：多以麻绳或皮革制成。

Seat surface: Usually made of flax ropes or leather.

脚踏：雕刻有几何形式的图案。

Footboard: Decorated with carved geometric patterns.

腿：两腿相交，可以开合折叠。

Legs: The two legs are intersected for folding and opening.

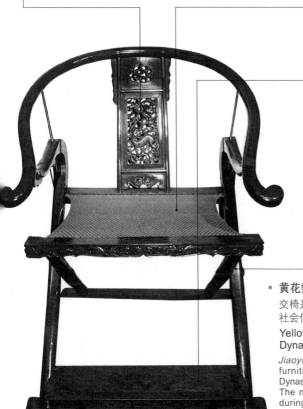

- **黄花梨木圆后背交椅（明）**

  交椅是一种折叠式椅子。宋元时期交椅是上层社会使用的高档家具，明清时期逐渐减少了。

  Yellow Rosewood Round-back *Jiaoyi* (Ming Dynasty, 1368-1644)

  Jiaoyi is a type of folding chair and high-range furniture used by the upper class of the Song Dynasty (960-1279) and Yuan Dynasty (1279-1368). The number of such chairs decreased gradually during the Ming Dynasty (1368-1644) and Qing Dynasty (1644-1911).

- 灯挂椅（明）

 灯挂椅是一种靠背椅，搭脑两端出挑，向上翘起，很像中国江南农村使用的竹制油盏灯的提梁，故而得名。灯挂椅整体多光素无雕饰，有装饰的灯挂椅也仅在靠背上雕饰简练精美的图案。整体由下向上略呈收势，给人以稳健、挺拔的视觉效果。

Lamp-handle Chair (Ming Dynasty, 1368-1644)

Lamp-handle chair is another type of chair with a back. The two tops of the horizontal beam of the back tilt upward like the handle of a bamboo lamp in rural southern China. Lamp-handle chairs usually have no carved decorations and those that do may only have simple carved patterns on the back. The shape of the chair narrows gradually from the bottom, giving it a stable and straight look.

- 皇宫圈椅（明）

 圈椅是扶手椅的一种变体形式。圈椅之名缘于靠背与扶手相连成圈形，靠背板向后凹曲，中心多有装饰图案。坐靠时，人的臂膀倚着圈形的扶手上，会感到十分舒适。

Imperial Round Chair (Ming Dynasty, 1368-1644)

The round chair is a variant of the armchair. Its name comes from its feature that the back and armrest are connected with a circle. The back leans backward and has decorations in the middle. When sitting in it, one can make use of the armrests for greater comfort.

- 酸枝木太师椅（清）

明清时期的太师椅没有固定式样，形体偏大，成排陈设在厅堂之上。太师椅的上部和下部为独立的两个部分，上部是屏风式的靠背和扶手，椅腿截面多为直角的方形，靠背和扶手也近于方形，整体粗犷厚重。

Blackwood *Taishi* Chair (Qing Dynasty, 1644-1911)

Taishi chairs of the Ming Dynasty (1368-1644) and Qing Dynasty (1644-1911) had no stereotype and were rather large. They were put in rows in the main hall. The upper and lower parts of the chair are independent. The upper part is the screen-like back and armrest. The cross-sections of legs are usually strict squares and the back and armrest are nearly square. The whole chair looks stately.

- 玫瑰椅（清）

玫瑰椅是为坐于书桌旁写作之用而设计的，故又名"文椅"。玫瑰椅的靠背和扶手与椅座均为垂直相交；靠背较低，与扶手高低相差不大。此椅低靠背，呈方形，靠背板透雕吉祥纹饰。座面上木材的自然纹理清晰，牙子雕饰简洁。

Rose Chair (Qing Dynasty, 1644-1911)

The rose chair is designed for sitting and writing at a desk and is also known as a study chair. Its back and armrests are vertical to the seat; the back is rather short, nearly at the same height as the armrest. The square back is decorated with openwork patterns of lucky images. The wood used for the seat has a smooth texture and the spandrel is simply ornamented.

- 官帽椅（清）

  官帽椅因其形似中国古代的官帽而得名，多置于厅堂明间的两侧，与茶几配套摆放，适合待客。

  Official Hat-shaped Armchair (Qing Dynasty, 1644-1911)

  The name comes from the chair's shape — it looks like the hat worn by Chinese bureaucrats in the old days. Official hat-shaped armchairs were usually placed along the two sides of the main hall to match tea tables for receiving guests.

- 明代的官帽

  The Hat Worn by Bureaucrats in Ming Dynasty (1368-1644)

- 榉木方凳（明）

  方凳指凳面正方形，面下有四足的凳子。

  Beech Square Stool (Ming Dynasty, 1368-1644)

  Square stools have a square seat and four legs.

- **梳背椅（明）**

  梳背椅是靠背椅的一种款式，因较宽的靠背是用细圆柱均匀排列而成，如同梳子，故名。

  Comb-back Chair (Ming Dynasty, 1368-1644)

  Comb-back chairs have backs made from thin sticks arranged in an even and orderly way like a comb.

- **酸枝木躺椅（清）**

  躺椅是一种靠背很高、又可大角度向后仰伸、椅座较长、带扶手的椅式，因人可仰躺在上面而得名。

  Blackwood Lounge (Qing Dynasty, 1644-1911)

  The lounge has a tall back that can be stretched backwards. It has a long seat and armrest for people to lie on, hence its name.

- 鼓式坐墩（明）

  坐墩是一种无靠背的小型坐具，圆形，腹部大，上下小，造型很像古代的鼓，可用草、藤、木、漆、瓷、石等材料制成。

  Drum-like Sitting Block (Ming Dynasty, 1368-1644)
  The sitting block is a small piece of furniture without a back. The round block has a large belly and the upper side is larger than the lower side, which makes it look like ancient drums. It may be made of grass, vines, wood, lacquer, porcelain, stone and other materials.

西番莲纹：西番莲纹是指从西洋传入的一种花卉纹饰，西番莲花色淡雅，自春至秋相继不绝。在家具中常作缠枝花纹，多作为边缘装饰。

Passionflower patterns were introduced from the West. The flowers have light and elegant colors and blossom from spring to autumn. Such patterns are commonly used in decorating the edges of furniture.

- 紫檀木圆凳（清）

  圆凳因坐面为圆形而得名，多带束腰，凳面为圆形、梅花形或海棠形。此凳的凳面为海棠形，光素平滑，有束腰，上饰以花卉纹。束腰与腿之间的牙子及腿足上浮雕西番莲纹，十分精美。

  Rosewood Round Stool (Qing Dynasty, 1644-1911)
  Round stools have round seats and are usually bundled beneath. The seat is in the shape of a circle, plum blossom or begonia. This stool has a begonia-shaped seat that is simple and smooth. It is bundled beneath and decorated with patterns. The spandrel and legs have delicate relief carvings of the passionflower.

- **黄花梨木滚凳（明）**

  滚凳是一种脚踏，具有保健功能，脚按滚凳能活动经络，有活血化瘀的功效，利于人体健康。

  Yellow Rosewood Rolling Stool (Ming Dynasty, 1368-1644)

  A rolling stool is a type of footboard that has health-promoting functions. Rolling one's feet on the stool can activate blood vessels and promote blood circulation, which is good for the health.

踏床：位于正面两足之间，上面钉有铜饰件，两端有插入足端卯眼中的圆轴，使踏床能被折起。

Footboard: The footboard is between the two front legs and decorated with copper ornaments. There are axles on both sides for drilling in mortises so that the footboard can be folded.

机面：多用绳索、丝绒或皮革条带等材料制成。

Seat surface: Usually made of ropes, velvet, leather strips or other materials.

机腿：用四根圆材制成，用来穿铆轴钉的断面呈方形，给人坚实之感。

Legs: Made up of four sticks. The cross-sections for drilling nails are square, giving it a stable appearance.

- **黄花梨木交机（明）**

  交机俗称"马扎"，是一种可折叠的方形坐具，由于携带、存放方便，千百年来一直被人们广泛使用。

  Yellow Rosewood Jiaowu (Ming Dynasty, 1368-1644)

  The Jiaowu, also called *"Mazha"*, is a square folding stool. It has been widely used for several hundred years because it is convenient to carry and stored.

传统靠背椅靠背的倾角和曲线

中国传统家具中，许多靠背椅的靠背都是弯曲的，其倾角和曲线设计的科学、合理，注重人体尺度，椅背倾角为100°，与今天国家公布的标准《常用家具尺寸》中的椅背尺寸的倾角97°—100°基本一致。这表明明代匠师在设计家具时，参考人体脊柱在自然状态时呈S形这一生理特征，将椅背做成S形曲线，与人体脊柱相适应，并根据人体休息时的必要后倾度，将靠背做成100°的背倾角，使人体后背与椅背接触充分，韧带和肌肉得到很好的休息，产生舒适感。

The back angle and curve of traditional chairs

In traditional Chinese furniture, many chairs have curved backs and the angle and curve are scientific and reasonable to be suitable for the scale of people's bodies. The back's angle is 100 degrees, which is basically in line with the 97-100 degrees in today's national Common Furniture Measurements. It shows that during the Ming Dynasty (1368-1644), when designing furniture, craftsmen took into consideration that the natural position of people's backbone is in the shape of an "S" and made the chair back into such a shape to adapt to the spine's natural position. They also made the back's angle 100 degrees so that one's back can fully and safely recline. In this way, one's ligaments and muscles may have a good rest and the sitter can feel very comfortable. horizontal architraves of different width, plus carved decorative motifs such as curling tendrils, lotus petals, rotary beads, Buddhist doors, animals and cloud and water, which is very decorative.

> 桌案类

桌案类家具是中国传统家具中品种最多的一类,人们常将它们并称。

桌是一种高型家具,常跟椅或凳配套使用。高型桌子逐渐普及使用,是从唐代开始的。发展到明清时期,桌的式样更多,有些桌和案的形状十分相似,一般认为桌比案略小,并且在形制结构上有两个显著特点:一是面板的长宽比不超过2:1,如果面板的长宽比超过此数值,则一般称为案;二是腿足安装在面板的四角处,称为"桌形结构"。明清桌有炕桌、酒桌、琴桌、棋桌、供桌、宴桌、圆桌、方桌、长方桌、半桌等形式。

案是一种案面为长方形,下有足的家具,多为木制。案在结构上的特点是,腿足不在面板的四角之

> Tables and Desks

The table and desk have the biggest variety of styles in traditional Chinese furniture and are usually mentioned together.

The table is tall furniture that is usually used together with a chair or stool. The wide use of the table began in the Tang Dynasty (618-907) and styles increased quickly during the Ming Dynasty (1368-1644) and Qing Dynasty (1644-1911). Some tables have very similar shapes as desks but usually they are smaller than desks. In addition, tables have two distinct features in their structure. Firstly, the length-width ratio is usually less than 2:1, otherwise, it is called a desk. Secondly, a table's legs are beneath the four corners of the table's surface. Tables of the Ming Dynasty (1368-1644) and Qing Dynasty (1644-1911) included the *Kang* table, wine table, Chinese lute table, chess table, worship table, banquet table, round table, square table, rectangle table and semi-table.

下，而是安装在案两侧向里收进的位置上，而且两侧的腿间嵌上雕刻多种图案的板心和各式券口。秦汉时期，案用来承载饮食用具，到了魏晋南北朝时期细分为食案、书画案、奏案、香案等。两宋时期，传统的矮形案除了用于床榻上的还保

The desk has a rectangular surface and feet and is usually made of wood. A structural feature of desks is that they have just two legs and these are not beneath the corners of the surface but further in. Moreover, various patterns and aprons (*Quankou*) are inlaid between the two legs as decorations. During the Qin Dynasty

- **黄花梨木炕桌（明）**

炕桌是在炕上、榻上使用的矮形家具。炕桌的尺寸不大，长、宽的比例约为3∶2，式样却很多。宫廷或官宦人家的炕桌制作十分考究，使用方式也相对固定，主要放于炕床一侧或坐榻中间。民间炕桌则讲究实用，制作比较古朴简单，使用也较灵活，白天放于炕上，晚上放于地上，夏季时还放于室外庭院中作地桌用。

Yellow Rosewood *Kang* Table (Ming Dynasty, 1368-1644)

A *Kang* table is short and used on a *Kang* or a couch. Its length-width ratio is about 3:2 and they come in various styles. Kang tables used by imperial or civil servant families were very delicate and had a specific use, mainly put on one side of the *Kang* or by the center of the couch. *Kang* tables used by common households highlighted their practicability and were simply made. The usage was rather flexible — they were put on the *Kang* during day time and on the ground at night and in the yard during summer to serve as a table.

留传统造型外，大部分逐渐消失，出现了富贵家庭中厅堂里陈设的高足案。明代时，案是用途最广的大型家具，主要有平头案、翘头案，大多陈设在大厅正中。

几也是桌案类家具的一种，是案面狭长、下有足的矮形家具，一般设于座侧，方便人们坐时依凭和搁置物件。

(221 B.C.-206 B.C.) and Han Dynasty (206 B.C.-220 A.D.), desks were used to contain food and divided into specialist styles such as food desks, study desks, musical desks and censer desks during the Jin Dynasty (265-420) and the Southern and Northern Dynasties (386-589). During the Northern Song Dynasty (960-1127) and Southern Song Dynasty (1127-1279), only traditional short desks that were used on beds kept their original style and tall desks began to emerge and were displayed in rich families' halls. Desks, including flat desks and tilting desks, became the most commonly used large furniture during the Ming Dynasty (1368-1644) and were usually placed in the main hall.

Ji is a type of table with a long and narrow surface and short legs. *Ji* is usually put beside the seat for people to lie against and put things on.

- 茶几（明）

茶几是专门用来摆设茶具的家具，通常带有霸王枨，有托泥装饰，玲珑精致。茶几与炕几有很多相似或通用之处，用途多与会客、宴饮有关。

Tea Table (Ming Dynasty, 1368-1644)

A tea table is used especially for the placing of tea sets. It is usually decorated with giant's arm brace and supporting sticks and is very delicate and exquisite. The tea table has many things in common with the *Kang Ji* and is usually used during guest receptions and banquets.

- **核桃木香几（清）**

  香几是宋代以后出现的一种摆放香炉用的高腿家具，一般形体高大，庄重大方。此几造型别致：几面方形，四周起拦水线；高束腰，嵌装透雕中国四大名著之一《西游记》人物故事图；足端上翻，亦透雕花纹；足下带方座，承托香几。

  Walnut incense stand (Qing Dynasty, 1644-1911)

  The incense stand appeared during the Song Dynasty (960-1279). It is a tall piece of furniture on which censers (incense burners) are placed. It is usually large and imposing. This stand has a delicate and solemn appearance. It has a square surface with water-barring skintle to prevent clothes from getting wet. The surface is bundled beneath where there are openwork images of people and a story from one of China's four classic novels — *Journey to the West*. Its feet roll upwards and are ornamented with openwork patterns. A square base is under the surface to support the stand.

- **竹节花几（清）**

  花几是摆放盆花的家具，可用于室内外，一般为方形、圆形、六角形等，有高有矮，成对使用。上品花几多选用名贵木材花梨木、紫檀木等制作，造型高雅，腿足设计精巧，几面常嵌大理石、玉、玛瑙、五彩瓷面等。

  Bamboo Flower Stand (Qing Dynasty, 1644-1911)

  The flower stand is used for potted flowers indoors. Usually it is square, round or hexagonal, either tall or short, and used in pairs. A high-end flower *Stand* is made of precious yellow rosewood or rosewood with delicate designs. The legs and feet are exquisitely designed and the surface is usually inlaid with marble, jade, agate or colored porcelain.

琴桌（清）

琴桌是抚琴时专用的桌子。琴桌形体不大，比一般桌略矮些，桌面似条桌，窄而长，桌四面饰有围板。桌膛由两层木板组成，其中留出透气孔，使桌子形成共鸣箱。清代琴桌制作讲究，透雕繁复，多作为陈设，置于厅堂，依墙而设，以示清雅。

Chinese lute Table (Qing Dynasty, 1644-1911)

Chinese lute table is made especially for playing the Chinese lute. It is not very large, slightly shorter than common tables and has a long and narrow surface. It has decorations around the surface and the lower part is made up of two layers of wood board. There are air holes on the board to form a resonance box. Such tables of the Qing Dynasty (1644-1911) were delicately manufactured with complex openwork decorations. The table was usually put in the main hall against the wall to create an elegant impression.

酒桌（明）

酒桌用于陈置酒肴，是一种形制较小的长方形案，比炕桌略大些，带有吊头，比一般的桌子要矮一些。有的桌面下设有双层隔板，用来放置酒具。两个侧腿之间则设双枨，以保证酒桌的结实。

Wine Table (Ming Dynasty, 1368-1644)

The wine table is used to display wine and dishes. It is square, has protruding ends and is a bit larger than a Kang table but shorter than the common table. Some have two-layer sheets to place a wine set. There are sticks between the two side legs to ensure that the table is stable and firm.

- 棋桌（近代）

棋桌是一种专用于弈棋的方桌或长桌。明清时期的棋桌设计巧妙，制作精美，桌面能活动，一般为双套面，个别的有三层。下棋时拿下一层桌面，便会露出棋盘，套面之下有暗屉，用来存放棋具。不用时，盖上桌面可当一般桌子使用。

Chess Table (Modern Times)

The chess table is a square or long table especially for playing chess. The chess table during the Ming Dynasty (1368-1644) and Qing Dynasty (1644-1911) was ingeniously designed and delicately manufactured. The surface can move and usually consists of two or even three layers. The first layer can be taken off to reveal the chessboard for playing, and there are drawers under the layer to store chess set. The surface is covered when there is no chess being played so it can serve as a common table.

- 圆柱式独腿圆桌（清）

圆桌是清代才开始流行的桌式，桌面为圆形，常由两张半圆桌拼成，也有制成独面的折叠圆桌和圆柱式独腿圆桌等。清代中期以后大圆桌流行，有的可围坐十几人至二十人。此桌造型优美，桌面下正中呈独腿圆柱式，圆柱上有托角花牙支撑桌面，下有站牙抵住圆柱，并和下面的踏脚相接，支撑稳固起桌面。上、下两节圆柱以圆孔和轴相套接，使桌面可自由转动。

Cylindrical One-leg Round Table (Qing Dynasty, 1644-1911)

Round tables became popular during the Qing Dynasty (1644-1911). The surface is round, usually made up of two semi-circular parts. Some can be folded and some are cylindrical one-legged ones. Large round tables began to be very popular during the middle Qing Dynasty; some can seat 10 to 20 people. This table has a beautiful cylindrical shape and only one leg. It has a spandrel on the top to support the surface and a standing spandrel at the bottom to support the cylinder, which is connected with the footboard to support the surface. The upper and lower parts of the desk are linked by a hole and an axle so that the surface can turn freely.

挡板：指在前后腿与横枨之间镶嵌的装饰性侧板。一般挡板用料较厚，使案腿更加稳固，其上常镂空雕刻各种吉祥纹饰，或用木条攒接成棂格形状，精巧空透，有很强的装饰性。此挡板透雕二龙戏珠图案，寓意吉祥。

Baffles: Decorative plates between the front and back legs and the horizontal sticks. Usually baffles are very thick so as to make the legs more stable and firm. They are decorated with hollowed-out auspicious patterns or connected by wood strips into the square shapes — delicate and elegant and very ornamental. This baffle is decorated with openwork patterns of two dragons playing with pearls, representing auspiciousness.

案面：呈长方形，中部平整光滑，两端向上翘起。

Surface: Rectangular, flat and smooth in the middle and tilting upwards at the two ends.

牙板：此牙板雕有葡萄云纹，寓意多子多福。

Spandrel: Decorated with grape and cloud patterns representing good luck and many children.

飞角：案面两端上翘的部分。

Fei Jiao: The tilting parts of the surface.

- **雕葡萄翘头案（清）**

  翘头案的案面两端装有翘头，设挡板，挡板多用较厚的木料，一般镂空雕刻精美的图案。翘头案多设在厅内中堂，也有在侧间使用，多放于窗前或山墙处，用来摆放花瓶或梳妆用具。

  Tilting Desk with Grape Carvings (Qing Dynasty, 1644-1911)

  The two ends of the surface tilt upward and the desk has baffles that are usually made of thick wood and decorated with beautiful hollowed-out patterns. Such desks are usually put in the main hall, some also in side rooms. They are used in front of windows or beside gables to hold vases or dressing sets.

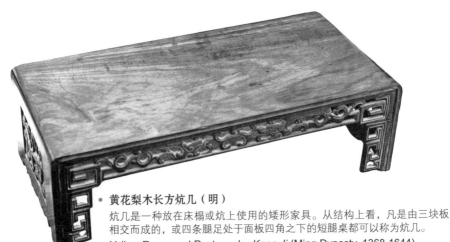

- **黄花梨木长方炕几（明）**

  炕几是一种放在床榻或炕上使用的矮形家具。从结构上看，凡是由三块板相交而成的，或四条腿足处于面板四角之下的短腿桌都可以称为炕几。

  Yellow Rosewood Rectangular *Kang Ji* (Ming Dynasty, 1368-1644)

  Kang Ji is short furniture used on a *Kang* or a couch. From the perspective of structure, all short tables that are made up of three intersecting plates or four legs beneath the four corners of the surface can be called *Kang Ji*.

- **雕三多平头案（清）**

  平头案是放在书房中用来写字作画的案，案面尺寸较宽大，不设抽屉，两侧加横枨，有的还有托泥装饰。三多纹是佛手、蟠桃、石榴合在一起的纹样，寓意多福、多寿、多子。

  Flat Desk with "Three-many" Patterns (Qing Dynasty, 1644-1911)

  A flat desk is put in the study room for writing and drawing on. It is usually large and has no drawers. Its two sides are decorated with horizontal sticks and supporting sticks. "Three-many" patterns refer to patterns combining citrons, peaches and guavas, representing much luck, long life and many children.

灵芝纹：灵芝被古人视为仙草，是祥瑞的征兆，因此常作为家具上的装饰纹样。桌案上的透雕灵芝纹，枝叶卷转，匀称妥帖。

Ganoderma patterns: Ganoderma was regarded as an immortal herb by ancient Chinese people and a symbol of auspiciousness. Therefore, it is usually used on furniture for decoration. The openwork ganoderma patterns on this table have rolling leaves and an even structure.

- **灵芝纹供桌（清）**

  供桌是中国古代年节时供奉祖先，或寺庙中用来陈设祭品时，放置壶、杯、盘等祭器的桌子。寺庙祠堂中所用供桌一般形制高大，有金漆或雕刻装饰，专放祭品。

  Worship Table with Ganoderma Patterns (Qing Dynasty, 1644-1911)

  Worship tables were used in ancient China to worship the ancestors or display sacrifices, kettles, cups and plates in temples. Worship tables used in temples especially for sacrifices are usually large and tall and decorated with gold paint or carvings.

苏州耦园内的家具摆设
Furnishing in Garden of Couple's Retreat (Ou Yuan), Suzhou

> 箱柜类

箱柜类家具主要指储藏物品，存放、搁置器物的箱、柜、橱和架格等。箱、柜的使用始于夏商时期，而橱到了汉代才处于初始阶段，且形体较高大。宋代以后，柜与箱区别明显，柜的形体逐渐变得高大，架格和橱也出现了多种形制。

一般来说，橱比柜小些，宽度大于高度，顶部采用面板结构，面板和门一样主要看面，既可当案来用，又可搁置物品。柜的体积高大，高度大于宽度，柜顶上没有面板结构，柜门是主要看面，双开或四开门，柜内装樘板数层。两扇柜门中间有立栓，柜门和立栓上装有铜饰件，可以上锁。还有一种具有柜和橱两种功能的家具，叫"柜

> Cabinets and Chests

This category of furniture mainly refers to chests, cabinets, closets and stands that are used to place and store stuff. Chests and cabinets started to be used in the Xia Dynasty (approx. 2070 B.C.-1600 B.C.) and Shang Dynasty (1600 B.C.-1046 B.C.) and the closet originated during the Han Dynasty (206 B.C.-220 A.D.). It was larger and taller than the former two. After the Song Dynasty (960-1279), the differences between cabinets and chests became prominent as cabinets were larger and taller. Stands and cabinets evolved into various styles.

Generally speaking, a closet is smaller than a cabinet and its width is larger than its length. Its top adopts the board structure so that it can be used both as a desk and for storage. The door and the board are the main parts. A cabinet is larger and the length exceeds the width. There is no board on the top. It has two doors forming

橱"，形体不大，高度相当于桌案，柜上的面板可作桌面使用。面板下安有抽屉，在抽屉下安有两扇对开的柜门，内装樘板，分为上下两层，门上有铜质饰件，可以上

the main part and several layers within the body. Between the two doors are bolts that are decorated with copper and can be used to lock the cabinet shut. Another type of furniture, combing the functions of a cabinet and a closet, is called a "closet cabinet". It is not as big or as high as a table. The board on the top can be used as a table surface. The drawers under the board have two doors. Inside the doors are two layers of boards and on the doors are copper decorations that can be used to hang a lock.

箱底：托座式箱底，正面浮雕双龙戏寿纹，祥瑞富贵。

Base: The supporting base is decorated with relief patterns of two dragons playing with the Chinese character "*Shou*", meaning auspiciousness and wealth.

拍子：箱子上主要的铜饰件，也有些衣箱用锁匙。

Hasp (*Paizi*): Main copper ornament on the chest. Some clothes chests use keys instead of *Paizi*.

提环：椭圆形提环设于箱子两侧，方便提携。

Lifting Handle (*Ti Huan*): *Ti Huan* is oval loops on the two sides of the chest to enable it to be lifted.

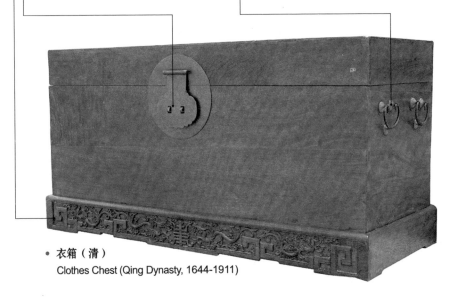

- 衣箱（清）
Clothes Chest (Qing Dynasty, 1644-1911)

- 大漆描金官皮箱（明）

官皮箱是一种旅行中用来贮物的小木箱子，体积不大，制作极其精美。最下面是底座，箱体上一般雕刻有喜庆吉祥图案。官皮箱用于盛装梳妆用具，也可用来装文具。

Guan leather Chest with Painting and Gold Decorations (Ming Dynasty, 1368-1644)

The *Guan* leather chest is a small wood chest used to contain possessions during a journey. It is small and delicately made. A base is at the very bottom and auspicious patterns are carved on the chest body. Such chests are used to store clothes and stationary.

- 榉木钱箱（近代）

钱箱是古代用来存放铜钱的木箱子，体积较大，重量也大，所用木料也较厚。为了取存方便，店铺中用来存放钱财的箱子也不一定做成箱式，有的做成柜式。这些钱箱虽然笨重，不易移动，却十分安全，又可兼作凳、床榻使用。

Beech Coin Chest (Modern Times)

A coin chest made of wood was used to keep copper coins. It is large, heavy and uses thick wood. To make it convenient to put in and take out money, the coin chests used in shops were sometimes made into cabinets. Such chests were big, heavy and hard to move but were very safe and served as chairs and beds.

• 冰箱（清）

中国古人在夏天也使用冰箱保存食物，不过古时的冰箱一般为木制，锡里，里面放有天然冰块（冬天把冰块储存在冰窖里供夏天使用），外部有铜箍，木盖上有镂空的钱式孔，下部有约一尺高的木座承托箱身。

Freezing Chest (Qing Dynasty, 1644-1911)

Ancient Chinese people used freezing chests to store food. Such chests were made of wood and painted with tin on the inner walls. There were natural ice cubes (stored in the icehouse during winter for use in the summer) inside and copper hoops outside. The wood cover had hollowed-out holes and the chest had a 30-centimeter-high wooden base to support the chest's body.

• 竹雕方形提盒（清）

提盒是带提梁的长方形箱盒，内有多层隔板，可放物品，有大、中、小三种规格，均为便于扛抬提行的盛具。大号提盒盛东西很多，需要两人穿杠抬行，中号提盒一人可挑两件，小的提盒可提在手中。提盒是盛具，古人郊游时携带馔肴酒食，或饭馆外送酒菜，店铺送货上门，文人盛放文具赶考等，都用提盒。

Bamboo Chest with a Handle (Qing Dynasty, 1644-1911)

This type of chest is square and has a handle. Inside the chest are multiple layers. Such chests have large, medium and small sizes that are all convenient to carry. The large-sized version can contain many things and it requires two people to lift it. Just one person can carry two medium-sized chests and the small-sized version can be held in hand. Such chests were used by people in the old days to carry wine and food during journeys, to send food to homes by restaurants, to send goods to doors from shops and to hold stationary by scholars taking the national exam.

- 酸枝木多宝格（清）

清代架格中以多宝格最为流行，多宝格又叫"博古格"，是用横、竖板将架格隔成大小不同、高低错落的多层小格，专门用来陈放文玩古器。

Blackwood *Duo Bao Ge* (Qing Dynasty, 1644-1911)

The *Duo Bao Ge* was the most popular furniture of its kind during the Qing Dynasty (1644-1911). It is also called a "*Bo Gu Ge*" and consists of multi-layer cases made of horizontal and vertical sticks. It is especially used for holding antiques.

- 药箱（清）

药箱即装药材的箱子，体积很小，箱体内设有十几个大小不同的小抽屉，可以存放不同的药品，便于搬动携带。

Drug Chest (Qing Dynasty, 1644-1911)

A drug chest, as its name suggests, is used to store drugs. It is small and has dozens of drawers of various sizes that can contain different drugs. It is easily carried.

后背：用攒斗工艺斗接处的花纹以四瓣枣花作心，将四根S形弯材接到花朵上，呈现出精美的波纹图案。

Back: Jujube flower patterns made using the *Zan Dou* technique. Four "S-shaped" bent sticks are connected by flowers, presenting delicate wave patterns.

抽屉：两个抽屉设在架的中部，高度和人胸际相当。上安铜吊牌拉手，可放置小物件。

Drawers: Two drawers are in the middle of the *Liang ge*, which comes up to one's chest and has copper hang tag. They can contain small items.

亮格：共两层，上面可放置书籍、古玩、器皿等物件。

Liang ge: Two layers with the upper one for books, antiques and containers.

- **黄花梨木透空后背架格（明）**

  架格是以立木为四足，用横板将空间分割成几层，用来陈设、存放物品的高形家具，是厅堂、书房之中常用的陈设。

  Yellow Rosewood *Jia Ge* (Ming Dynasty, 1368-1644)

  A *Jia Ge* has four legs and is divided into several layers by horizontal sticks. Such tall furniture is used to place things on and store belongings. It is the major piece of furniture in the main hall and the study.

- **五斗橱（近代）**

  五斗橱的特点是柜高不超过120厘米；柜顶上有面板，可摆放物品；看面设五个大抽屉，抽屉面多用细木镶嵌或用瘿木作贴面装饰；通常不设柜门，即使设柜门也很小；柜框边缘浮雕装饰图案。

  Wudou Closet (Modern Times)

  The *Wudou* closet is no higher than 120 centimeters. There is a board on the top on which items can be placed and five large drawers to the front that are decorated with thin wood inlay or cecidium. Usually it has no doors or, if any, only very small ones. The edges of the closet frame are decorated with relief patterns.

- **柜橱（清）**

  柜橱是由闷户橱演变而来的一种橱子，抽屉下没有闷仓。其抽屉以下空间设计成一个尽可能大的柜体，正面安柜门，流行于清代中晚期。

  Cabinet Closet (Qing Dynasty, 1644-1911)

  The cabinet closet evolved from the pantry cabinet and has no pantries beneath it. The under part of the drawers is made into a cabinet as large as possible and it has doors in the front. It was popular during the middle and late Qing Dynasty.

亮格：指柜子上没有门的开敞式隔层，可陈设古玩，放置器物。有的亮格用隔板分出隔层。前面设有栏杆，两侧用立柱固定，中间透雕花纹，后背和两侧均空敞。

Shelf (*Liangge*): Open shelves without doors. They can be used to hold antiques and other items. Some shelves have baffles to create interlayers. The shelf has rails in front and is fixed by sticks on both sides. There are openwork patterns in the middle and no rails at the back or along the two sides.

抽屉：设于亮格和柜之间，上有铜拉环，可存放小件物品。

Drawers: Between the shelf and the cabinet with copper loops. Used to contain small items.

面叶：此面叶呈长条状，因此又称"面条"。

Plate (*Mianye*): These *Mianye* are long strips and so are also called strip (*Mian Tiao*).

- **榉木亮格柜（明）**

亮格柜是架格和柜子的组合，常是架格在上，柜子在下，用于陈设和收藏。架格有一层，也有双层或多层的，与人肩齐高或稍高些。亮格柜一柜两用，多放在厅堂或书房，极富文人气息。

Beech Shelf Cabinet (Ming Dynasty, 1368-1644)

The shelf-cabinet is a mixture of a stand and a cabinet, usually with the stand on the top and the cabinet on the bottom. It is used to place and store stuff. The shelf may have one, two or more layers and is as high as one's shoulder. Such double-functional cabinets are usually put in the main hall or the study room to add a studious atmosphere.

> 屏风类

屏风是中国传统家具中极具特色的一种，有分隔室内空间、挡风、屏蔽视线、装饰等诸多用途。屏风起源于西周，形制较矮，至汉代时使用广泛，经常与茵席、床榻结合使用。唐代，随着高足家具的逐渐流行，屏风渐渐高大起来。屏风有较好的室内美化功能，各种场合都可摆设屏风。式样以座屏和曲屏居多，而且书画名家在屏风上题诗作画在唐代成为一种时尚。宋代屏风有直立板式、多扇曲屏等式，制作精美，除挡风、遮障功能外，更多是作为一种文化的载体，多设在主人会客之处。明代屏风主要用于室内空间隔断和装饰，主要有座屏和曲屏两大类，基本上沿用了前代的样式，做工装饰更加精美华

> Screens

A screen is one of the most typical traditional items of Chinese furniture, serving purposes such as separating room space, blocking draughts and obscuring views as well as acting as decoration. Screens originated during the Western *Zhou* Dynasty (1046 B.C.-771 B.C.) and were very short. During the Han Dynasty (206 B.C.-220 A.D.), screens became popular and widely used together with mats, beds and couches. Screens became larger and taller during the Tang Dynasty (618-907) as tall furniture became more popular. They were used at various occasions for their visual beauty. Stand with pedestal and folding screens were major types and it became a fashion that famous painters and calligraphers wrote poems and drew pictures on them. Screens of the Song Dynasty (960-1279)

丽，或为雕刻，或为镶嵌，或为绘画，或为书法。明代后期出现一种挂在墙壁上的挂屏，用于装饰。清代屏风品种繁多，形体雄大，屏心常镶大理石、玻璃等饰品。

included standing ones and folding ones that were delicately manufactured. They served as a carrier of spirits and culture besides the purposes mentioned above. They were usually placed where the host received guests. During the Ming Dynasty

渔樵耕读图案：用中国古代农耕社会四种职业或身份——渔夫、樵夫、农夫与书生——作为家具的雕刻图案，常作为官宦退隐之后生活的一种象征。

Yu qiao geng du pattern: It refers to four important occupations in ancient Chinese agrarian society — fishing, woodcutting, farming and scholarship. The pattern is widely used on furniture as decoration, symbolizing the desires of bureaucrats after retiring from their jobs.

- 象牙雕座屏（清）

座屏又叫"插屏"，插在屏座之上，做工精美，为陈设欣赏品。

Lvory Stand with Pedestal (Qing Dynasty, 1644-1911)

Stands with pedestal are also called *Cha Ping*, meaning that the screen is on a base. The manufacturing is very delicate and the screen is for display and appreciation.

屏框：多用较轻质的木材做成，便于摆放和折叠收藏。

Frame: Usually made of light wood so that it was convenient to place, fold and store.

屏心：由八扇屏面组成，用轻质木材做成，上有清末书法家赵之谦篆字书法，苍秀雄浑，清高脱俗。

Screens: Made up of eight fans with light wood. The strong and powerful calligraphy of Zhao Zhiqian, a famous calligrapher of the late Qing Dynasty (1644-1911), was applied to the screens.

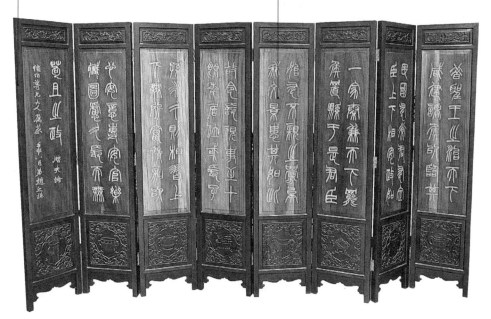

- 曲屏（清）

  曲屏又名"围屏"、"折屏"，是一种可以折叠的多扇屏风，落地摆放，采用攒框做法，扇与扇之间用铜合页相连，可以随时拆开，有二扇、四扇、六扇、八扇、十二扇等样式。曲屏为临时性陈设，摆放位置随意。摆放时，扇与扇之间形成一定的角度便可摆立在地上。为了营造某种氛围，或体现地位的高低，经常用曲屏来重新划分室内的空间，以增强每个空间的相对独立性，满足使用者的要求。曲屏还常围在床榻旁，既可以遮挡，又可以凭靠。

Folding Screen (Qing Dynasty, 1644-1911)

The folding screen is also called a *"Wei Ping"* and *"Zhe Ping"*. It can be folded and put on the ground. The *Zan Kuang* technique is adopted and the fans are connected with copper *Heye*. Fans can be opened when necessary and include two, four, six, eight and 12 fan types. Folding screens are occasional furniture that can be placed anywhere. They can stand on the ground with certain angles made by the fans. Such screens are usually used to reshape the room to create a certain atmosphere or to show one's social status. In this way, each space of the room is more independent and certain needs are met. Folding screens are also put around beds or couches to block out draughts or for lying against.

• 百宝嵌竹石纹挂屏

挂屏出现在明代晚期,是一种纯粹的装饰品和陈设品,悬挂在墙上代替卷轴画。常见的装饰技法有百宝嵌、嵌瓷、嵌玉、嵌珐琅、玻璃油画等,丰富多彩。

Treasure Inlay Bamboo and Stone Pattern Hanging Screen

The hanging screen appeared during the late Ming Dynasty (1368-1644) and was a purely decorative display. They were hung on the wall to replace scroll paintings. Common decorations include treasure inlay, inlaid porcelain, inlaid jade, inlaid enamel and oil painting on glass.

(1368-1644), screens were mainly used to separate space and decorate the room and consisted of stand with pedestal and folding screens that basically kept the appearance of previous periods. The manufacturing and decoration were more exquisite with carving, inlaid material, painting or calligraphic ornamentation. The practice of hanging screens on the wall emerged during the late Ming Dynasty (1368-1644), mainly as a decorative device. Screens of the Qing Dynasty (1644-1911) came in various styles and were much larger. They were decorated with marble and glass.

传统家具的保养

传统家具是在日常使用中传世的,其时代特征——包浆,也是在日常使用中形成的。因此,对传统家具,不可将其像纯观赏性装饰品那样供起来。

传统家具是采用严密的榫卯结构装配起来的,年深日久会因自重、承重过大、受拉力过大、受热、受潮等原因,出现受力的榫卯结构松动的现象。因此在日常使用中避免上述因素的发生,便是最好的日常保养。

另外,传统家具上忌放置音响设备。因为音响设备使用时产生的振动,会引发谐振而造成家具的损坏。避免直接在古家具上放一些发热物品,以免破坏家具表面的包浆。还要避免与硬物摩擦,以免损伤漆面和木面纹理,如放置瓷器铜器等装饰物品时要特别小心,最好是垫上一层软布,轻拿轻放,以免破坏漆膜。

Maintenance of traditional furniture

Traditional furniture is passed down through generations in daily usage and its feature of the times —*"Bao Jiang"* — is also formed in the same process. Therefore, traditional furniture cannot be regarded as pure decorative display.

Traditional furniture is manufactured using tightly fastened tenon and mortise structures. They may become looser over time because of weight, tension, heat or moisture. Therefore, avoiding the above-mentioned factors is the best way to maintain traditional furniture.

Stereo equipment should not be place on traditional furniture because the vibration caused by such equipment can damage it. Hot items should not be put directly on traditional furniture in case that the *"Bao Jiang"* is destroyed. Friction with hard materials must also be avoided to prevent damage to paintings and texture. For instance, special caution must be taken when placing porcelain and copper decorations. It is better to use a soft cloth under them to avoid damage to any paintings.

- 现代家庭中的传统家具摆设

  传统家具不仅是一种实用品，还能营造情调或文化氛围，供人欣赏、享受，这种功能主要是靠家具的陈设来实现。

  Traditional Furniture Layout in Modern Houses

  Traditional furniture is not only practical but also creates a certain cultural atmosphere. People can appreciate its beauty as well as its functionality.

• 现代家庭中的传统家具摆设
Modern Pieces of Traditional Furniture

> 支架类

支架类家具既是立体支撑的承物器，又是一种美化室内环境的装饰品，包括灯架、衣架、盆架、帽架、镜架、鸟架、笔架、鱼缸架、

> Stands

Stands can support other articles. They also serve as decorations that beautify the interior environment. They include lamp stands, coat stands, basin stands,

- **六柱脸盆架（清）**

盆架是用于放置洗脸盆的家具，分高、矮两种。高盆架是巾架和盆架的结合，多为六足，最里面的两足加高成为巾架，最上面的横杆是搭挂洗面巾用的，两端出挑，多雕有云头或凤首。矮盆架腿足等高，上端和下端各装一组横枨，专用于放脸盆。

Six-legged Basin Stand (Qing Dynasty, 1644-1911)

Basin stands are used for supporting copper basins. They are categorized into tall basin stands and short basin stands. A tall basin stand includes a towel hanger. It is usually six-legged, with the innermost two legs elongated. The towel hanger consists of the two elongated legs and a horizontal rail used for hanging towels. The ends of the horizontal rail are usually carved cloud clusters or phoenix heads. A short basin has legs of the same length. At each end of the legs there is a group of horizontal rails used for supporting the basins.

- **黄花梨木凤纹衣架（明）**

  衣架是用于搭衣服的木架，有支架和横杆，通常放在卧室床榻旁边或进门的一侧，并与床、桌、椅等室内家具在风格和尺寸上协调一致。

  Yellow Rosewood Coat Stand with Phoenix Patterns (Ming Dynasty, 1368-1644)

  Coat stands are made of wood and used to hang coats. A coat stand consists of supporting legs and horizontal rails. It is usually placed beside the bed or the door and should be in tune with the size and style of other furniture such as the bed, table and chairs.

- **屏式穿衣镜（民国）**

  屏式穿衣镜出现在清代中期，广州从外国进口大尺寸镀水银厚玻璃镜，安装在座屏中成为穿衣镜。镀水银玻璃镜在当时是珍稀名贵之物，只有宫廷王府才用得起。

  Screen-like Full-length Mirror (the Republic of China)

  The screen-like full-length mirror first emerged in the mid-Qing Dynasty (1644-1911), when Guangzhou imported large and thick mercury-coated glass and used them in seat screens. Mercury-coated mirrors were quite rare and precious at that time and could only be afforded by royal families.

兵器架和乐器架等品种，其中衣架、盆架、灯架和镜架是传统家具中较为典型的架具品种。

hat stands, mirror stands, bird stands, pen stands, aquarium stands, weapon stands and musical instrument stands. Coat stands, basin stands, lamp stands and mirror stands are typical stands in traditional furniture.

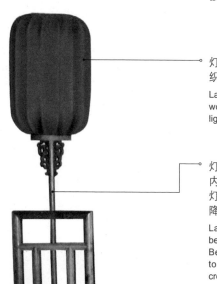

灯罩：多用竹或木材制成架子，外面糊丝织物，用以聚光、防风雨。

Lamp shade: Frames (mostly made of bamboo or wood) covered in silk fabrics used to soften the light and as protection during bad weather.

灯杆：圆形灯杆插入底座横梁上的圆孔内，孔旁有木榫，下端和活动横木相连，灯杆和横木能顺着底座内的槽口上下升降，升降的高度由木榫固定。

Lamp post: The cylinder-shaped lamp post can be put into the hole of the crossbeam at the base. Beside the hole is a wood plug which is linked to the movable crossbeam. The lamp post and crossbeam can move along the notch in the base and the height is fixed by the wood plug.

- 升降式灯架（清）

灯架是专用来承放油灯或蜡烛的家具，多为木制，摆放随意，有装饰作用。汉代以前灯架多为铜制，较低矮，常置于席上或几案之上使用。明清时期灯架可分为固定式、升降式和悬挂式三类。其中，升降式灯架可以根据实际需要来调整其高度，多在喜庆吉日用于厅堂照明。

Elevator-type Lamp Stand (Qing Dynasty, 1644-1911)

Lamp stands hold lamps or candles. They are commonly made of wood and can be put anywhere to decorate the house. Before the Han Dynasty (206 B.C.-220 A.D.), most lamp stands were shorter and made of copper. They were usually put on the bed or table. By the Ming Dynasty (1368-1644) and Qing Dynasty (1644-1911), the lamp stands can be divided into three styles: the fixed type, the elevator type and the suspended type. This elevator type's height can be adjusted and they are mostly used on festive occasions to light the hall.

苏州网师园五味书屋的家具摆设
Furniture Layout of Wuwei Study in Suzhou the Master-of-Nets Garden Park

《中国红》出版编辑委员会

主　　任　　王亚非

副 主 任　　田海明　林清发

编　　委　　（以姓氏笔划为序）

　　　　　　王亚非　　毛白鸽　　田海明　　包云鸠

　　　　　　孙建君　　任耕耘　　吕　军　　吕品田

　　　　　　吴　鹏　　杜国新　　林清发　　赵国华

　　　　　　徐　雯　　涂卫中　　唐元明　　韩　进

　　　　　　蒋一谈

执行编委　　任耕耘　　蒋一谈

中国红系列

Chinese Red Series

传统手工艺 Traditional Chinese Crafts	风筝 Kites	面具 Masks
中国色彩 Colorful China	盆景 Bonsai	鼻烟壶 Snuff Bottles
刺绣 Chinese Embroidery	景泰蓝 Cloisonné	四大发明 Four Great Inventions of Ancient China
中国禅 Zen	泥塑 Clay Sculpture	丝绸之路 The Silk Road
棋艺 Art of Chesses	面塑 Dough Figuring	汉字 Chinese Characters
宋词 Ci-Poems of the Song Dynasty	大运河 Grand Canal	中国木偶艺术 Chinese Puppet Arts
茶马古道 Ancient Tea-Horse Road	中国历史名城 Historical Cities	古代兵书 Ancient Book on the Art of War
中国名湖 Famous Lakes in China	中国结 Chinese Knots	道教文化 Daoism Culture in China
北京中轴线 The central axis in Beijing	兵马俑 Terracotta Army	古代交通 Ancient Traffic
帝王陵寝 Imperial Mausoleum	皮影 Folk Shadow Play	古代壁画 Ancient Chinese Mural Painting
中华传统美德 Chinese Traditional Virtues	中国古代帝王 Emperors of China	古代衡器 Ancient Weighing Apparatus
中国姓氏 Chinese Surnames	中国陶器 Chinese Pottery Ware	24节气 The Twenty-four Solar Terms
传统家具 Chinese Furniture	中国漆器 Chinese Lacquer Articles	中国名泉 Famous Springs in China
中国名山 Renowned Chinese Mountains	中国名寺 Famous Temples in China	长江黄河 Yangtze River and Yellow River
中国染织 Chinese Dyeing and Weaving	中国石窟 Grottoes in China	传统杂技 Traditional Acrobatic Arts
武术 Chinese Martial Arts	中国古桥 Ancient Bridges in China	中国婚俗 Marriage Customs in China
民间玩具 Folk Toys	中国古塔 Ancient Pagodas in China	匾额对联 Inscribed Tablets and Couplets
古代教育 Education in Ancient China	中国民居 Traditional Civil Residents	中国建筑装饰 Chinese Architectural Decoration
中国神话传说 Chinese Mythology and Legends	民间戏曲 Traditional Folk Operas	十二生肖 The Twelve Animals Represent Years
古代游戏 Recreational Games in Ancient China	中国灯彩 Colorful Chinese Lanterns and Lamps	古代佩饰 Ornaments Wore by Ancient Chinese
四大名著 Four Masterpieces of Chinese Fiction	诸子百家 Traditional Philosophers and Ideologists	文房清供 Stationery and Bibelot in Ancient Studies
古代科技 Ancient China's Science and Technology	中国牌坊 Chinese Decorated and Memorial Archways	中国祥禽瑞兽 Auspicious Beasts and Fowls in Chinese Culture
金银器 Gold and Silver Wares	中国茶艺 Chinese Tea Appreciation Ceremony	
竹木牙角器 Art Crafts Make of Bamboo, Wood, Ivory and Horn	秦砖汉瓦 Brick of Qin Dynasty and Tile of Han Dynasty